AF358681

Advances in the Diagnosis and Management of Temporomandibular Joint Diseases 2.0

Advances in the Diagnosis and Management of Temporomandibular Joint Diseases 2.0

Editor

Luis Eduardo Almeida

Basel • Beijing • Wuhan • Barcelona • Belgrade • Novi Sad • Cluj • Manchester

Editor
Luis Eduardo Almeida
Surgical Sciences - Oral and
Maxillofacial Surgery
Marquette University School
of Dentistry
Milwaukee
United States

Editorial Office
MDPI
St. Alban-Anlage 66
4052 Basel, Switzerland

This is a reprint of articles from the Special Issue published online in the open access journal *Diagnostics* (ISSN 2075-4418) (available at: www.mdpi.com/journal/diagnostics/special_issues/ Temporomandibular_joint).

For citation purposes, cite each article independently as indicated on the article page online and as indicated below:

Lastname, A.A.; Lastname, B.B. Article Title. *Journal Name* **Year**, *Volume Number*, Page Range.

ISBN 978-3-7258-0400-9 (Hbk)
ISBN 978-3-7258-0399-6 (PDF)
doi.org/10.3390/books978-3-7258-0399-6

Contents

About the Editor

Luis Eduardo Almeida

Dr. Luis Eduardo Almeida is a passionate professional with an illustrious academic journey. He embarked on his quest for knowledge by obtaining a Doctor in Dental Surgery (DDS) degree (1989–1993) from the esteemed Federal University of Parana State, UFPR, Brasil. His dedication led him to earn a Certificate in Oral and Maxillofacial Surgery (1995–1998) from the same institution.

He continued his academic pursuits with fervor, achieving a Fellowship in Oral and Maxillofacial Surgery (1998–1999) at Northwestern University, Chicago, USA, followed by a Residency Program in the same field (2001–2003) at the same prestigious institution. He further enhanced his expertise by completing a Master of Science in Health Sciences (2004–2006) and a Ph.D. in Health Sciences (2008–2013) from Pontifícia Universidade Católica do Paraná, PUC-PR, Brasil.

Currently, Dr. Luis Eduardo Almeida radiates enthusiasm as a beacon of inspiration at Marquette University—School of Dentistry. As a Clinical Associate Professor in Surgical Sciences—Oral & Maxillofacial and Oral Surgery—he infuses his lectures with boundless energy and real-world insights. His dynamic leadership as Director of the Predoctoral Program in Oral & Maxillofacial and Oral Surgery (since 2018) ignites a passion for excellence within his students.

Board-certified by the Brazilian Dental Board and the Wisconsin Dental Board, Dr. Luis Eduardo Almeida is a trailblazer in his field. His contagious enthusiasm extends beyond academia, as evidenced by his enriching experience as an Oral & Maxillofacial Surgeon in private practice (2003-2013).

In every endeavor, Dr. Luis Eduardo Almeida embodies the spirit of enthusiasm, innovation, and excellence, leaving an indelible mark on the world of dentistry.

Preface

Welcome to the second edition of "Advances in the Diagnosis and Management of Temporomandibular Joint Diseases". This volume represents a collaborative effort to push the boundaries of our understanding of temporomandibular joint (TMJ) disorders and refine the approaches to their diagnosis and management.

The field of TMJ disorders is dynamic and multifaceted, encompassing various disciplines such as dentistry, rheumatology, neurology, and surgery. This reprint is a testament to the diversity and innovation within the field, showcasing a collection of research that explores novel diagnostic techniques, innovative interventions, and the complex interplay between TMJ disorders and various medical conditions.

As they navigate through these pages, readers will encounter a rich tapestry of insights, each contributing to the collective knowledge of TMJ disorders. The studies presented reflect the interdisciplinary collaboration that defines modern healthcare, with contributions from clinicians, researchers, and technologists working together to advance the field.

From investigations into the impact of rheumatic diseases on the TMJ to explorations of postural changes in patients with dystonia following TMJ intervention, the papers within this reprint shed light on the intricacies of TMJ disorders. Cross-sectional studies, pilot interventions, and arthroscopic explorations offer a comprehensive view of the current state of research, providing valuable information for professionals and researchers alike.

The preoperative considerations of temporomandibular disorders in orthognathic surgery, the diagnostic and surgical role of arthroscopy in synovial chondromatosis, and the evaluation of TMJ in patients with Parkinson's disease are among the diverse topics covered. Each paper, in its own way, contributes to the mosaic of knowledge that defines the current landscape of TMJ research.

We express our gratitude to the contributors for their dedication to advancing the understanding and management of temporomandibular joint diseases. It is our sincere hope that this compilation serves as a catalyst for further collaboration, exploration, and innovation in the pursuit of improved outcomes for those affected by TMJ disorders.

Luis Eduardo Almeida
Editor

Article

Temporomandibular Disorders at the Preoperative Time of Orthognathic Surgery

Alice Vanzela Miotto, Danielle Veiga Bonotto, Jessica Sarha Cavalheiro Silva, Juliana Feltrin De Souza, Aline Monise Sebastiani * and Rafaela Scariot

Department of Stomatology, School of Dentistry, Federal University of Paraná, 623 Prefeito Lothário Meissner Avenue, Curitiba 80210-170, PR, Brazil; alicevmiotto@gmail.com (A.V.M.); jessicasarahcavalheiro@hotmail.com (J.S.C.S.); julianafeltrin@ufpr.br (J.F.D.S.); rafaela_scariot@yahoo.com.br (R.S.)
* Correspondence: sebastiani.aline@gmail.com

Abstract: Individuals seeking orthodontic treatment combined with orthognathic surgery (OS) have a high prevalence of temporomandibular disorders (TMDs), but the relationship between TMD diagnoses and dentofacial deformities (DFDs) is still controversial. Therefore, this cross-sectional study with a comparison group aimed to analyze the association between dentofacial deformities and TMDs. Methodology: Eighty patients undergoing OS were consecutively selected from the stomatology department of the Federal University of Paraná between July 2021 and July 2022. Forty patients who would undergo OS composed the group of participants with DFD, and forty who received other types of attention and did not present changes in the dental bone bases formed the group without DFDs (DFDs and no DFDs groups). The groups were matched for sex, age, and self-reported ethnicity. The diagnostic criteria for TMDs (DC/TMDs) were used to diagnose TMD based on the Axis I criteria. The psychosocial aspects, oral behaviors in wakefulness, and sleep bruxism were evaluated through the Axis II criteria. The data were analyzed with a 5% significance level. Results: The presence of DFDs was significantly associated with arthralgia ($p = 0.01$). The other types of TMDs were not associated with DFDs. Comorbidities, habits, and psychosocial variables were not associated with DFDs at a level of 0.05 ($p > 0.05$). In analyzing the participants with arthralgia, the ones with this condition presented higher frequencies of sleep bruxism ($p = 0.046$). Conclusions: Participants with DFDs presented a significantly higher frequency of arthralgia when compared to no DFDs ones. Sleep bruxism was associated with the occurrence of joint TMDs in these participants.

Keywords: temporomandibular disorders; dentofacial deformity; sleep bruxism

Citation: Miotto, A.V.; Bonotto, D.V.; Silva, J.S.C.; De Souza, J.F.; Sebastiani, A.M.; Scariot, R. Temporomandibular Disorders at the Preoperative Time of Orthognathic Surgery. *Diagnostics* **2023**, *13*, 2922. https://doi.org/10.3390/diagnostics13182922

Academic Editor: Luis Eduardo Almeida

Received: 27 July 2023
Revised: 5 September 2023
Accepted: 7 September 2023
Published: 12 September 2023

1. Introduction

Dentofacial deformities (DFD) correspond to dental malocclusion that is associated with skeletal pattern alterations [1]. Such deformities can be minimal, as in a slight projection of the chin, or extreme, as in a severe vertical maxillary excess or a hemifacial microsomia. The involvement may be in one or two bases of the bone in the vertical, horizontal, and transverse planes, both in isolation and in combination, thus causing different types of deformities. The main DFDs are skeletal class II, skeletal class III, skeletal biprotrusion, a skeletal anterior open bite, and a skeletal bilateral posterior crossbite [2].

The moderate and severe cases of DFDs require a combined treatment between orthodontics and orthognathic surgery (OS). Orthognathic surgery consists of repositioning the bone bases through osteotomies to align and correct the jaw's position, thereby improving function and facial aesthetics. Pain management in orthognathic surgery is essential to enhance recovery, reduce hospital stay, and improve the whole experience of the patient [3]. The postoperative period of this surgery involves a difficult recovery process, with restrictions on food, difficulties in opening the mouth, significant swelling of the face, and frequent joint discomfort [4].

Among the reasons for performing OS, aesthetic purposes are frequently reported, as are functional improvements, including complaints related to temporomandibular dysfunction [5–7]. Although current concepts no longer consider occlusion to play a central role in the occurrence of TMDs, when it comes to dentofacial deformities, several studies [8,9] suggest a high prevalence of this condition in these individuals, especially in groups seeking surgical treatment. It is also important to note that many of these patients often have emotional problems, including anxiety and depression, due to the negative impact of the deformity [10,11].

Temporomandibular disorders (TMDs) encompass heterogeneous conditions involving the masticatory muscles, the temporomandibular joint (TMJ), or both, as well as their associated structures. Several studies have investigated TMDs in orthosurgical patients [12,13]; however, the way the deformity impacts specific TMD diagnoses still needs to be determined. There is controversy among the studies and a scarcity of papers that evaluate comparison groups using valid tools. The Research Diagnostic Criteria for Temporomandibular Disorders (RDC/TMD) [14] instrument followed the biopsychosocial model of TMD assessment and classification, which consists of Axis I (physical diagnoses) and Axis II (psychosocial aspects) criteria. It was updated in 2014 for the Diagnostic Criteria for TMDs (DC/TMDs) and recently validated for the Portuguese language in Brazil [15].

This study aimed to compare orthosurgical patients with dentofacial deformity (DFD group) and individuals without dentofacial deformity (no DFD group) regarding TMD diagnoses.

2. Material and Methods

This is an observational cross-sectional study with a comparison group, which was developed in the Oral and Maxillofacial Surgery and Traumatology Service (OMSTS) facilities and the dental clinics of the Department of Stomatology at Federal University of Paraná (UFPR), Jardim Botânico campus, located in Curitiba in south of Brazil. The study went on for 36 months. Calibration within the examiners was performed using the kappa coefficient. To assess the inter-rater reliability, the Kappa coefficient was used (k:095/95% CI).

3. Ethical Aspects

The longitudinal study was approved by the Research Ethics Committee under protocol CAAE number: 52207821.9.0000.0102. In addition, the study followed all the recommendations of the Declaration of Helsinki [16] regarding research with humans. Individuals were invited to participate in the study and received information about the objectives and justifications for the research through an informed consent form (ICF). They received information about the benefits and risks to which they would be exposed. They were also aware that the treatment would continue, regardless of their refusal to participate in the research, and that they had the freedom to discontinue participation at any time. Individuals who consented to participate in the study signed the ICF and were included. Only the researchers had access to the questionnaires to ensure the confidentiality of the data.

4. Sample

The study included a total of 80 participants. The Open Epi Software, updated version 6 April 2013, calculated the sample size. The calculation was based on a previous study in which a prevalence of 31% was found in the general adult population [17]. For the group no DFDs, a prevalence of 63.8% was found in patients with DFDs who would undergo OS [12]. Thus, the calculation was performed with a bilateral confidence interval of 95% and a power of 80%, with a ratio of the no DFDs group to the DFDs group of 1:1. Among the 80 selected participants, 40 formed the group with DFDs, and 40 formed the group without DFDs.

The sample selection was performed as follows: for the group with DFDs, all individuals who would undergo OS by the OMSTS Service of UFPR and who met the inclusion criteria were invited to participate. The participants were recruited when they went through the clinical examination stage before OS. The inclusion criteria for the group with DFDs were the following: have dentofacial deformities not associated with cleft lip and palate or syndromes, requiring treatment through OS, over 18 years of age, accepted to participate in the research, and having signed the ICF.

For each individual selected to the DFDs group, we looked for an individual matching the sex, age, and self-reported race in another dentistry clinic at the UFPR to maintain a more homogeneous sample and to eliminate potential biases. These individuals were approached in the first evaluation in other dental clinics of the UFPR. The individuals who agreed to participate and who met the inclusion and exclusion criteria were selected. The inclusion criteria for the group without DFDs comprised patients over 18 years of age, who were undergoing maintenance dental treatment, such as cleaning, bleaching, replacement of restorations, who signed the ICF, and who did not present dentofacial deformities. An experienced professional conducted a dentofacial deformity analysis.

The exclusion criteria for both groups were the presence of previous surgeries in the cervicofacial region, cognitive and neurological alterations, the diagnosis of arthritis and arthrosis in other joints, tooth pain, oral pathologies, and patients who used cyclobenza-prine in the week of the evaluation.

5. Data Collection

Demographic data were collected from all survey participants, such as sex, age, and self-reported race; they were divided into whites and nonwhites due to the n and low number of other races. In addition, the data were collected on drug use and the presence of comorbidities (fibromyalgia and arthritis/arthrosis), and data were collected on patients' habits such as smoking (smoker and nonsmoker) and the amount of coffee intake (less than 3 cups and more than 3 cups per day).

For the analysis of DFD types, an experienced surgeon evaluated the participant's profile and occlusion, as well as classified the face profiles into three categories: I, II, or III. Profile I is a straight profile, and profile II presents a negative step between the maxilla and mandible corresponding to patients with mandibular retrognathism. Profile III presents a positive step between the maxilla and mandible, thereby indicating anteroposterior maxillary deficiency, mandibular prognathism, or both [18]. They were also evaluated to consider mandibular asymmetry (higher than 4 mm) and vertical facial patterns: an anterior open bite, vertical maxillary excess, and vertical maxillary deficiency.

All participants were diagnosed with TMDs through the DC/TMDs. The tool was used by trained and calibrated examiners. This tool consists of two axes: Axis I, which includes the information collected on the physical examination, and Axis II, which features emotional aspects. Both axes were used in this research [14].

Axis I consists of three questionnaires: TMD pain screening, DC/TMD symptom questionnaire, and TMD physical examination. To diagnose TMDs in patients, the DC/TMD procedures involved gathering the patient's medical history, performing a physical exam that included muscle palpation (applying 1.0 kg pressure) and joint palpation (applying 0.5 kg pressure), assessing the presence of symptoms during mandibular function, and measuring the maximum opening of the jaw both with and without pain. Based on the collected findings, the tool identified two main categories of physical diagnoses (muscular and joint). It generated a decision flowchart and a table of diagnostic criteria to assist in the diagnostic process; painful TMD conditions in the muscles were classified as myalgia, local myalgia, myofascial pain, myofascial pain with spreading, and myofascial pain with referred pain; Axis I also classified headaches attributed to TMDs as present or absent (Figure 1). Regarding temporomandibular joint disorders, they can be structural or joint pain per se (arthralgia), and both diagnoses are performed independently on each side of the (TMJ. Structural disorders are classified as disc displacement with reduction, disc displacement with reduction and intermittent locking, disc displacement without reduction and without limitation of the opening, disc displacement without reduction and with limitation of the opening, and degenerative joint disease (Figure 2). As for arthralgia, it was considered present or absent with respect to joint pain. The degenerative joint disease was diagnosed through the symptom of crackling during mandibular function without complementary imaging.

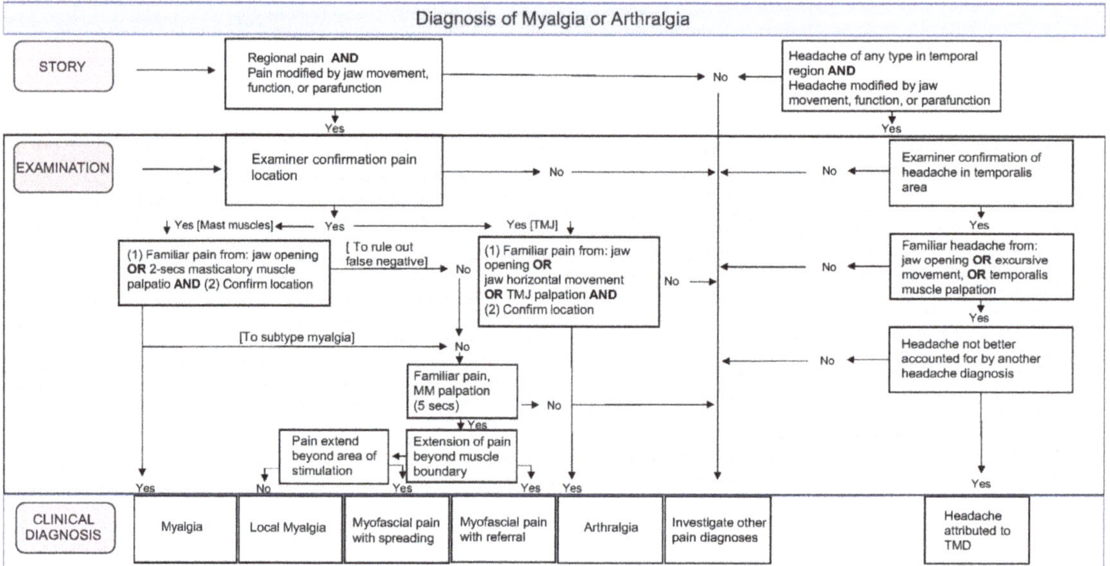

Figure 1. Flowchart of the criteria followed to myofascial pain, arthralgia and TMD Headache diagnose.

Axis II incorporates behavioral instruments regarding pain, the psychological state, and psychosocial and behavioral functioning. For Axis II, the tools used were the Patient Health Questionnaire (PHQ-15), the Generalized Anxiety Disorder (GAD-7), the human body pain drawing to identify pain points in and beyond the face, and the Oral Behaviors Checklist (OBC), which was used to diagnose awake and sleep bruxism. The OBC is a self-evaluation tool with 21 questions in which the patient answers the weekly and monthly frequency of oral and parafunctional behaviors. According to the sum of the scores, the result is classified as no oral behaviors for scores up to four points, mild from four to twelve points, moderate from thirteen to nineteen points, and severe with more than twenty points. Based on the OBC data, combined with the clinical evaluation, the awake bruxism was classified as absent, infrequent, frequent, or very frequent according to the answers to the OBC questionnaire of the DC/TMDs.

Sleep bruxism was classified as absent, possible, or probable. It was considered possible when it had a positive response in the OBC questionnaire and probable when, in addition to a positive response in the OBC, it presented at least one of the clinical signs: dental wear; marks on the soft tissue such as jugal mucus and tongue and/or muscle fatigue upon awakening; pain in the masseter and temporal muscle palpation; and/or masseter hypertrophy. [19] The PHQ-15 comprises 15 questions about physical symptoms such as nausea, dizziness, and body aches. The overall health conditions and the physical symptoms are classified according to severity level, from mild to severe. The instrument called GAD-7 consists of seven items to evaluate the symptoms of generalized anxiety disorder. The maximum total score is 21, where zero means the absence of anxiety, 5–9 means a mild degree, 10–14 indicates a moderate degree, and 15–21 means a severe degree of anxiety. Pain drawings are a self-applied tool represented by a drawing of the whole body and face in which the patient makes markings where they identify the pain.

The score is the sum of the number of markings. A pilot study to verify the methodology and applicability of the questionnaires was conducted with 10 patients with DFDs who met the inclusion and exclusion criteria of the present study and who would undergo OS at the Oral and Maxillofacial Surgery and Traumatology Service (OMSTS) of the UFPR in Paraná, Brazil. There were no intercurrences during the pilot study, and, from that, the methodology was then implemented, and these ten patients were already included in the sample.

Based on Diagnostic Criteria for Temporomandibular Disorders (DC/TMD): Diagnostic Decision Tree

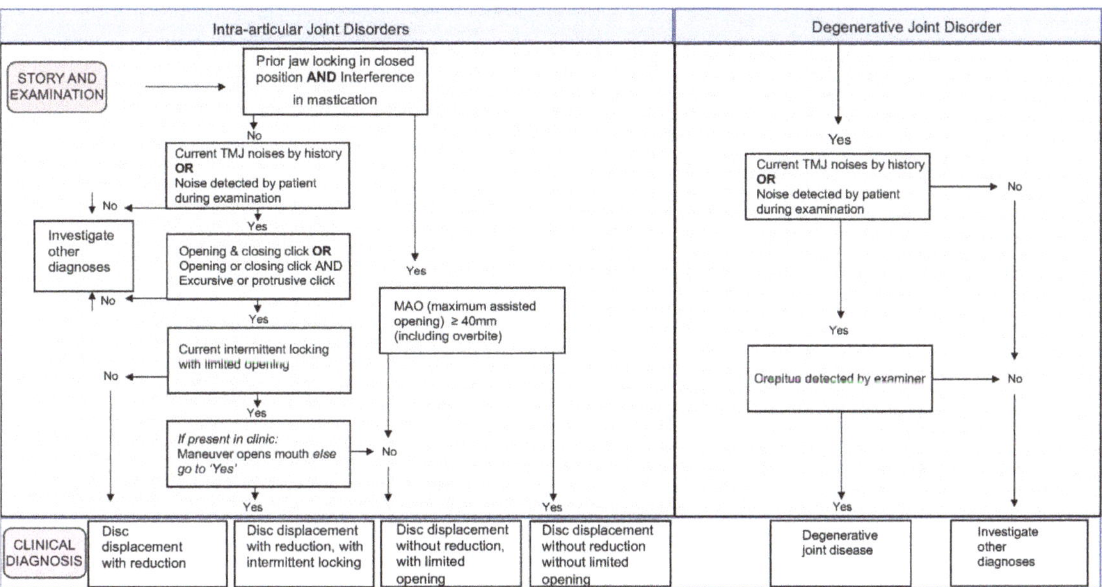

Figure 2. Flowchart of the criteria followed to temporomandibular joint disorders diagnose.

6. Data Analysis

The individuals were diagnosed according to the present or absence of the types of TMDs. Similarly, structural TMJ disorders were grouped as absent or present, thereby considering the sides of the TMJ (right and left) individually. Our study considered the presence of arthralgia, regardless of the affected side. As for the Axis II, the participants were classified as 1—without anxiety and mild anxiety—or 2—moderate and severe anxiety (GAD-7). Regarding the PHQ-15, the participants were classified as 1—without symptoms and mild symptoms—or 2—moderate and severe symptoms.

Regarding the OBC, parafunctional habits were dichotomized according to 1—absent and minor—or 2—present and very present. Wake bruxism was also dichotomized into 1—frequent and infrequent—or 2—frequent and very frequent.

The results obtained were submitted to descriptive and inferential statistical analysis. The categorical variables of the DC/TMDs between the groups were compared using the chi-square test or the Fisher exact test. For the numerical variables, the normality condition was evaluated by the Kolmogorov–Smirnov test. The comparison of numerical variables with non-normal distribution between groups was performed using the Mann–Whitney test. The variables of awake bruxism and sleep bruxism were compared with smoking and coffee habits in the general sample using the chi-square test or the Fisher exact test.

The level of significance adopted was 5%. The data were analyzed using the software *SPSS Statistics* v. 24 (IBM Corporation, Armonk, NY, USA).

7. Results

The sample was primarily composed of women, with 52 of them (65%) and 28 men (35%) being equally distributed between groups. The median age was 30 (19–61). Regarding the ethnicity, 65 participants (81.25%) were white, and 15 were nonwhite (18.75%). The groups were matched for sex ($p = 1.0$), age ($p = 0.823$), and self-reported race ($p = 1.0$).

There was no association between the variables awake bruxism and sleep bruxism with the variables smoking and drinking more than three cups of coffee in the general sample ($p > 0.05$).

Table 1 shows that the groups' presented homogeneity between some covariables, such as comorbidities and habits.

Concerning the DFDs group, 7 participants presented the facial profile I, 12 had the facial profile II, and 21 had the facial profile III. A total of ten participants presented asymmetry. Regarding the vertical alteration patterns of the face, six participants presented vertical excess, six presented vertical deficiency, and nine presented an open bite.

Table 1. The association of habits and comorbidities between the DFDs and no DFDs groups.

Variables		DFD *n* (%)	No-DFD *n* (%)	*p-Value*
Antidepressants Medication	No	37 (92.5)	36 (90.0)	1.000
	Yes	3 (7.5)	4 (10.0)	
Comorbidity	No	37 (92.5)	36 (90.0)	0.899
	Fibromyalgia	1 (2.5)	1 (2.5)	
	Gastroesophageal reflux	2 (5.0)	3 (7.5)	
Smokers	No	34 (85.0)	39 (97.5)	0.108
	Yes	6 (15.0)	1 (2.5)	
Coffee intake	0 to 3 cups	24 (60.0)	29 (72.5)	0.344
	>3 cups	16 (40.0)	11 (27.5)	

Note: Fisher exact test or chi-square test independent samples with a significance value of 0.05.

Table 2 shows the association between the diagnosis of TMDs and the groups. The presence of most TMD diagnoses was not associated with TMDs. However, individuals with DFDs presented a significantly higher frequency of arthralgia than those without DFDs ($p = 0.01$).

In analyzing the 18 participants who experienced joint pain (arthralgia) in the DFDs group, it was verified that 13 (72%) presented bilateral arthralgia, while 5 (28%) presented unilateral arthralgia. Also, 12 individuals (67%) presented disc displacement with reduction associated, with 4 of them presenting unilaterally, and in 8 (44.4%) of them it was present on both sides of the TMJ. In the group no DFDs, the four participants with arthralgia had no other joint disorders associated.

In terms of diagnosing disc displacement, our analysis found that, despite presenting similar results between the groups, 12 of them had painful symptoms related to their disc displacement, while 10 did not experience pain. Degenerative diseases in the TMJ were identified in one patient from the group with DFDs and in two from the group without DFDs. No patient had a disc displacement without reduction.

There was no difference in the maximum mouth opening between groups with and without DFDs. The median pain-free mouth opening value in the DFDs group was 43 mm (22–63 mm), and, in the group without DFDs, it was 45 mm (20–77 mm) ($p = 1.0$). The median mouth opening with pain was 48 mm (33–65 mm) in the DFDs group and 45 mm (30–78 mm) in the no DFDs group ($p = 0.117$).

Table 2. Comparison of TMD diagnoses between DFDs and no DFDs groups.

Diagnoses		DFD *n* (%)	No-DFD *n* (%)	*p*-Value
Myalgia	Absent	18 (45.0)	16 (40.0)	0.821
	Present	22 (55.0)	24 (60.0)	
Left DDwR	Absent	27 (67.5)	28 (70.0)	1.000
	Disc displacement with reduction	13 (32.5)	12 (30.0)	
Right DDwR	Absent	20 (50.0)	29 (72.5)	0.066
	Disc displacement with reduction	20 (50.0)	11 (27.5)	
Arthralgia	Absent	22 (55.0)	36 (90.0)	**0.010**
	Present	18 (45.0)	4 (10.0)	
TMD Headache	Absent	30 (75.0)	29 (72.5)	1.000
	Present	10 (25.0)	11 (27.5)	

Note: Chi-square test of independent samples with a significance value of 0.05. Bold values indicate statistical significance. Acronyms: DDwR—disc displacement with reduction.

Regarding the variables of Axis II of the DC/TMDs, there were no differences between the groups ($p > 0.05$) in anxiety levels (GAD-7), in physical symptoms (PHQ-15), and in pain in and beyond the face (Table 3). When considering the score for the pain points beyond the face on the pain drawing, there was also no difference between the groups; in the group with DFDs, the median pain score was 2 (0–19), and in the no DFDs group, it was 2 (0–12) ($p = 0.262$).

Table 3. Association between the Axis II DC/TMD variables between the groups.

Variables		DFD *n* (%)	No-DFD *n* (%)	*p*-Value
Physical symptoms	Absent and mild	15 (37.5)	19 (47.5)	0.498
	Moderate and severe	25 (62.5)	21 (52.5)	
Anxiety	Absent and mild	22 (55.0)	29 (72.5)	0.162
	Moderate and severe	18 (45.0)	11 (27.5)	
Pain in the face	Absent	15 (37.5)	19 (47.5)	0.176
	Present	25 (62.5)	19 (47.5)	
Pain Beyond the Face	Absent	16 (40.0)	21 (47.5)	0.370
	Present	24 (60.0)	19 (47.5)	

Note: Chi-square test of independent samples with a significance value of 0.05.

The variables of the subjects' self-reported race, age, facial profile, and asymmetry variables were unrelated to arthralgia within the DFDs group ($p > 0.05$). Regarding sex, women had a higher prevalence of joint pain ($p > 0.001$).

Table 4 shows the comparison between the participants with and without arthralgia (within the group with DFDs) in relation to the other variables of Axes I and II of the DC/TMDs and the parafunctional habits. Individuals with arthralgia presented a significantly higher frequency of sleep bruxism. The prevalence of probable sleep bruxism was 72% in individuals with joint pain compared to 59% in the group without joint pain ($p = 0.046$).

Table 4. Comparison of DC/TMD variables between participants with and without joint pain within the DFDs group.

Variables		No Arthralgia (22) *n* (%)	With Arthralgia (18) *n* (%)	*p*-Value
Myalgia	No	13 (59.0)	5 (27.0)	0.062
	Yes	9 (41.0)	13 (72.0)	

Table 4. *Cont.*

Variables		No Arthralgia (22) n (%)	With Arthralgia (18) n (%)	p-Value
Articular Disorder Left	No DDwR	18 (81.0)	9 (50.0)	0.341
	With DDwR	4 (18.0)	9 (50.0)	
Articular Disorder Right	No DDwR	13 (59.0)	7 (38.0)	0.046
	With DDwR	9 (41.0)	11 (61.0)	
Pain Beyond the Face	No	9 (41.0)	7 (38.0)	1000
	Yes	13 (59.0)	11 (61.0)	
Physical symptoms	Absent and mild	11 (50.0)	4 (22)	0.104
	Moderate and severe	11 (50.0)	14 (77)	
Anxiety	Absent and mild	14 (63.0)	8 (44)	0.225
	Moderateand severe	8 (36.0)	10 (55.5)	
OBC	None/light	5 (22.0)	3 (16)	0.339
	Moderate–severe	17 (77.0)	15 (83)	
	Absent	3 (13%)	5 (27)	
	Possible	6 (27%)	0 (0.0)	
	Probable	13 (59%)	13 (72)	
Wake Bruxism	Absent; uncommon	13 (59%)	8 (44)	0.525
	Common; very common	9 (41%)	10 (55.5)	
	Yes	4 (18%)	2 (11)	
Coffee intake	0 to 3 cups	13 (59%)	11 (61)	1.00
	>3 cups	9 (40%)	7 (38)	

Note: Fisher exact test and chi-square test of independent samples with a significance value of 0.05. Bold values indicate statistical significance.

8. Discussion

The diagnosis of TMDs is complex, mainly due to their different diagnostic methods and their multifactorial etiologies. Regarding the occlusal factor, after years of debate about the role of occlusal characteristics as causal or risk factors for TMDs, a low relevance for dental occlusion and the interarcaded relationship is currently attributed [13]. However, regarding DFDs, previous studies suggest a high prevalence of TMDs in this population, and a higher prevalence of pain and depression have also been reported compared to patients without DFDs [9,20].

This study focused on patients seeking OS, because it is necessarily understandable that temporomandibular dysfunctions are experienced in this population for the proper management and care of the patients before surgery. If patients undergo surgery with a previous pain, they will have more challenges in dealing with the postoperative period in functional and psychological terms. Also, long-lasting pain leads to changes in the central nervous system, thus causing central sensitization and increased postoperative pain sensitivity [21].

A recent systematic review showed that patients undergoing orthodontic surgical treatment have a higher incidence of TMDs when compared to a control population [9]. However, TMD subdiagnosis was not categorized in this systematic study, which only took into account the existence or absence of TMDs. Given the wide range of symptoms and treatment options available for TMDs, we do not believe that TMDs should be evaluated as a single diagnosis. Also, only two of the six studies that were a part of this systematic review used the RDC/TMD criteria to identify TMDs. Thus, to our knowledge, no previous studies have compared the diagnoses of DC/TMDs among patients with and without DFDs in the Brazilian population having used the DC/TMDs, which is the most accepted instrument for TMD diagnosis nowadays, thus emphasizing the importance of this study.

The main finding in this study was that surgical patients with DFDs present a higher prevalence of arthralgia compared to a control population, thus corroborating another study [22]. Arthralgia is a type of TMD that is associated with peripheral etiological factors such as parafunction and joint overload (which can occur during sleep or wakefulness) [23]. The control of this condition encompasses a combination of noninvasive therapies, including patient education, self-care, cognitive behavioral therapy, physiotherapy, pharmacotherapy, and occlusal devices. When conservative therapies are not effective, minimally invasive therapies such as the intraarticular injection (IAI) of hyaluronic acid (HA) or corticosteroid (CS), arthrocentesis, or arthroscopy could be useful, but all these therapies should be associated with overload control [24].

The relationship between bruxism and the symptoms of temporomandibular disorders is deeply discussed in the literature, due to the complexity of the etiology and diagnosis of both conditions [25,26]. In the present study, a significant difference was found concerning probable sleep bruxism, which was more prevalent in the group with joint pain. Thus, we suggest that the joint overload caused by this condition contributes to the development of TMDs. However, the control of the sleep bruxism and consequently of the arthralgia could be very challenging during preoperative orthodontic preparation, which ends up making it impossible to use interocclusal devices. Thus, new treatment protocols should be investigated for these patients.

In our study, we did not found any association between awake or sleep bruxism and smoking or drink coffee habits. However, is important observe that we had just seven smokers and three individuals that drank more than three cups of coffee in the general sample. So, it is possible that we did not find an association with these variables due to few individuals presenting these habits. According Bertazzo et al., more than eight small cups could be considered a risk factor for bruxism. Thus, we should not put away these habits as possible risk factors of bruxism [27].

Although there was no difference in the incidence of muscle TMDs between the groups, this study's findings revealed that both groups had a high prevalence of the condition (55–60%). This could be associated with the period of high stress and poor sleep quality in the population over the data collection period, which was during the pandemic (2020–2021), when there was a significant increase in these symptoms reported by the population. The result of a higher prevalence may be due to the increased incidence of muscle TMDs occurring in the pandemic period compared to previous years [28], thus resulting in a higher prevalence than studies before this period [29,30].

It is also important to highlight that other studies analyzing specific subtypes of deformities found some different findings. For example, a systematic review suggested that class II skeletal profiles and hyper-divergent growth patterns were likely associated with an increased frequency of TMJ disc displacement and degenerative disorders. Another study that used cone beam tomography and the RDC/TMDs to diagnose TMDs found more prevalent bone changes in patients with the class II skeletal malocclusion [25]. Also, a study comparing the prevalence of TMDs in patients with dentofacial deformities associated with class III malocclusion found that it was similar to patients without dentofacial deformities [31]. Thus, it is necessary to consider a limitation of the present study in that the type of dentofacial deformity in each individual was not classified.

This study also investigated psychosocial variables between groups and their relationships with arthralgia. Although we found no associations in this study, this could also be due to having a small and restricted sample. Temporomandibular disorders are complex conditions, and their interrelationships should be seen through pain models, thereby inserting the biopsychosocial perspective in the evaluations. Thus, these variables should continue to be investigated in other studies.

Thus, we believe that the most important limitation of this study is the sample size (DFDs = 40 and no DFDs = 40), given that important associations could be found with a larger n. It was also impossible to segment the sample according to the type of deformity. Another constraint was that bruxism was identified without polysomnography, which made it impossible to deliver a definite diagnosis according to the international consensus on bruxism [19]. Also, while the clinic holds the authority for TMD diagnosis as per the DC/TMD guidelines, and the signs and symptoms show high accuracy for specific diagnoses, it is possible that the disc displacement without reduction (DDwoR) without a limited opening went undiagnosed in this study due to the absence of complementary exams, such as magnetic resonance imaging (MRI) of the TMJ.

Therefore, we suggest further prospective cohort studies with larger samples to better investigate our findings, thereby focusing on treatment options for these patients. We also recommend future studies to investigate additional factors that may contribute to TMDs in this population.

Finally, it is important to point out that understanding the profiles of these patients and their functional problems before OS can positively influence the outcome of ortho-surgical treatment, which is still a wildly neglected factor by surgeons. The prior identification of patients with TMDs, parafunction, or both, should imply a treatment plan mainly incorporating cognitive–behavioral approaches to help patients understand their need to maintain relaxed the masticatory muscles.

9. Conclusions

The prevalence of arthralgia was higher in orthosurgical patients with dentofacial deformity when compared to individuals without dentofacial deformity. Sleep bruxism was associated with the occurrence of joint TMDs in these patients. Detecting these conditions and carrying out adequate management before surgery can ensure a better prognosis. Thus, further studies should investigate the additional factors that may contribute to TMDs in this population and determine new protocols of preoperative management.

Author Contributions: Methodology, R.S.; Software, J.S.C.S.; Formal analysis, J.F.D.S.; Data curation, D.V.B.; Writing—original draft, A.V.M.; Supervision, A.M.S. All authors have read and agreed to the published version of the manuscript.

Funding: This research received no external funding.

Institutional Review Board Statement: The study was conducted in accordance with the Declaration of Helsinki, and approved by the Research Ethics Committee under protocol CAAE number: 52207821.9.0000.0102.

Informed Consent Statement: Informed consent was obtained from all subjects involved in the study.

Data Availability Statement: The data presented in this study are available on request from the corresponding author. The data are not publicly available due to ethical restrictions.

Conflicts of Interest: The authors declare no conflict of interest.

References

1. Ferraz, O.; Rossi, R.R.P.; Fabbro, C.D.; Bittencourt, L.R.A.; Tufik, S. Cirurgia ortognática: Avanço maxilomandibular para tratamento da síndrome da apnéia obstrutiva do sono (SAOS). *Ortho Sci. Orthod. Sci Pract.* **2011**, 875–882.
2. Fish, L.C.; Epker, B.N.; Sullivan, C.R. Orthognathic surgery: The correction of dentofacial deformities. *J. Oral Maxillofac. Surg. Off. J. Am. Assoc. Oral Maxillofac. Surg.* **1993**, *51* (Suppl. S1), 28–41. [CrossRef]
3. Alyahya, A.; Aldubayan, A.; Swennen, G.R.J.; Al-Moraissi, E. Effectiveness of different protocols to reduce postoperative pain following orthognathic surgery: A systematic review and meta-analysis. *Br. J. Oral Maxillofac. Surg.* **2022**, *60*, e1–e10. [CrossRef]
4. Pereira, J.R.; Vicentini, G.; Meger, M.N.; Rebellato, N.L.B.; Scariot, R.; Sebastiani, A.M. Impact factors on oral health-related quality of life in patients requiring orthognathic surgery. *RSBO* **2020**, *17*, 106–111. [CrossRef]
5. Soh, C.L.; Narayanan, V. Quality of life assessment in patients with dentofacial deformity undergoing orthognathic surgery—A systematic review. *Int. J. Oral Maxillofac. Surg.* **2013**, *42*, 974–980. [CrossRef]
6. Finlay, P.M.; Moos, S.F.; Atkinson, J.M. Orthognathic surgery: Patient expectations; psychological profile and satisfaction with outcome. *Br. J. Oral Maxillofac. Surg.* **1995**, *33*, 9–14. [CrossRef]
7. Stirling, J.; Latchford, G.; Morris, D.O.; Kindelan, J.; Spencer, R.J.; Bekker, H.L. Elective orthognathic treatment decision making a survey of patient reasons and experiences. *J. Orthod.* **2007**, *34*, 113–127. [CrossRef] [PubMed]
8. Nicot, R.; Chung, K.; Vieira, A.R.; Raoul, G.; Ferri, J.; Sciote, J.J. Condyle modeling stability, craniofacial asymmetry and ACTN3 genotypes: Contribution to TMD prevalence in a cohort of dentofacial deformities. *PLoS ONE* **2020**, *15*, e0236425. [CrossRef] [PubMed]

9. Al-Moraissi, E.A.; Perez, D.; Ellis, E., 3rd. Do patients with malocclusion have a higher prevalence of temporomandibular disorders than controls both before and after orthognathic surgery? A systematic review and meta-analysis. *J. Craniomaxillofac. Surg.* **2017**, *45*, 1716–1723. [CrossRef] [PubMed]
10. Sebastiani, A.M.; Dos Santos, K.M.; Cavalcante, R.C.; Pivetta Petinati, M.F.; Signorini, L.; Antunes, L.A.A.; Rebellato, N.L.B.; Küchler, E.C.; Scariot, R. Depression, temporomandibular disorders, and genetic polymorphisms in IL6 impact on oral health-related quality of life in patients requiring orthognathic surgery. *Qual Life Res.* **2020**, *29*, 3315–3323. [CrossRef]
11. Okeson, J.P. *Orofacial Pain. Guidelines for Assessment, Diagnosis and Management*; Chicago, Quintessence Publishing: Chicago, IL, USA.
12. Sebastiani, A.M.; de Lucas Corso, P.F.C.; Bonotto, D.; de Souza, J.F.; da Costa, D.J.; Scariot, R.; Rebellato, N.L.B. Does orthognathic surgery improve myofascial pain in individuals with skeletal class III? One-year follow-up. *Oral Surg. Oral Med. Oral Pathol. Oral Radiol.* **2018**, *126*, 322–330. [CrossRef] [PubMed]
13. Manfredini, D.; Segù, M.; Arveda, N.; Lombardo, L.; Siciliani, G.; Rossi, A.; Guarda-Nardini, L. Temporomandibular Joint Disorders in Patients With Different Facial Morphology. A Systematic Review of the Literature. *J. Oral Maxillofac. Surg.* **2016**, *74*, 29–46. [CrossRef]
14. Schiffman, E.; Ohrbach, R.; Truelove, E.; Look, J.; Anderson, G.; Goulet, J.P.; List, T.; Svensson, P.; Gonzalez, Y.; Lobbezoo, F.; et al. Diagnostic Criteria for Temporomandibular Disorders (DC/TMD) for Clinical and Research Applications: Recommendations of the International RDC/TMD Consortium Network* and Orofacial Pain Special Interest Group†. *J. Oral Facial Pain Headache* **2014**, *28*, 6–27. [CrossRef] [PubMed]
15. Ohrbach, R.; Gonzalez, Y.; List, T.; Michelotti, A.; Schiffman, E. Diagnostic Criteria for Temporomandibular Disorders (DC/TMD) Clinical Examination Protocol. Available online: www.rdc-tmdinternational.org (accessed on 2 June 2013).
16. Gandevia, B.; Tovell, A. Declaration of helsinki. *Med. J. Aust.* **1964**, *2*, 320–321. [PubMed]
17. Valesan, L.F.; Da-Cas, C.D.; Réus, J.C.; Denardin, A.C.S.; Garanhani, R.R.; Bonotto, D.; Januzzi, E.; de Souza, B.D.M. Prevalence of temporomandibular joint disorders: A systematic review and meta-analysis. *Clin. Oral Investig.* **2021**, *25*, 441–453. [CrossRef]
18. Queiroz, G.V.; Rino Neto, J.; Paiva, J.B.D.; Capelozza Filho, L. Analysis of reliability, accuracy, sensitivity and predictive value of a subjective method to classify facial pattern in adults. *Dent. Press J. Orthod.* **2016**, *21*, 58–66. [CrossRef]
19. Lobbezoo, F.; Ahlberg, J.; Raphael, K.G.; Wetselaar, P.; Glaros, A.G.; Kato, T.; Santiago, V.; Winocur, E.; De Laat, A.; De Leeuw, R.; et al. International consensus on the assessment of bruxism: Report of a work in progress. *J. Oral Rehabil.* **2018**, *45*, 837–844. [CrossRef]
20. Sebastiani, A.M.; Gerber, J.T.; Bergamaschi, I.P.; Petinati, M.F.; Meger, M.N.; Costa, D.J.D.; Brancher, J.A.; Küchler, E.C.; Scariot, R. Individuals requiring orthognathic surgery have more depression and pain than controls. *Braz. Oral Res.* **2021**, *35*, e091. [CrossRef]
21. McCarberg, B.; Peppin, J. Pain Pathways and Nervous System Plasticity: Learning and Memory in Pain. *Pain Med.* **2019**, *20*, 2421–2437. [CrossRef]
22. Krisjane, Z.; Urtane, I.; Krumina, G.; Neimane, L.; Ragovska, I. The prevalence of TMJ osteoarthritis in asymptomatic patients with dentofacial deformities: A cone-beam CT study. *Int. J. Oral Maxillofac. Surg.* **2012**, *41*, 690. [CrossRef]
23. List, T.; Jensen, R.H. Temporomandibular disorders: Old ideas and new concepts. *Cephalalgia* **2017**, *37*, 692–704. [CrossRef] [PubMed]
24. Al-Moraissi, E.A.; Wolford, L.M.; Ellis, E., 3rd; Neff, A. The hierarchy of different treatments for arthrogenous temporomandibular disorders: A network meta-analysis of randomized clinical trials. *J. Craniomaxillofac. Surg.* **2020**, *48*, 9–23. [CrossRef]
25. Svensson, P.; Jadidi, F.; Arima, T.; Baad-Hansen, L.; Sessle, B.J. Relationships between craniofacial pain and bruxism. *J. Oral Rehabil.* **2008**, *35*, 524–547. [CrossRef]
26. Hang, G.J.; LeResche, L.; Critchlow, C.W.; Martin, M.D.; Drangsholt, M.T. Risk factors for diagnostic subgroups of painful temporomandibular disorders (TMD). *J. Dent. Res.* **2002**, *81*, 284–288. [CrossRef]
27. Bertazzo-Silveira, E.; Kruger, C.M.; Porto De Toledo, I.; Porporatti, A.L.; Dick, B.; Flores-Mir, C.; De Luca Canto, G. Association between sleep bruxism and alcohol, caffeine, tobacco, and drug abuse: A systematic review. *J. Am. Dent. Assoc.* **2016**, *147*, 859–866.e4. [CrossRef] [PubMed]
28. Slade, G.D.; Bair, E.; Greenspan, J.D.; Dubner, R.; Fillingim, R.B.; Diatchenko, L.; Maixner, W.; Knott, C.; Ohrbach, R. Signs and symptoms of first-onset TMD and sociodemographic predictors of its development: The OPPERA prospective cohort study. *J. Pain* **2013**, *14* (Suppl. 12), T20–T32.e3. [CrossRef]
29. Fernández-de-las-Penas, C.; Svensson, P. Myofascial Temporomandibular Disorder. *Curr. Rheumatol. Rev.* **2016**, *12*, 40–54. [CrossRef] [PubMed]
30. Janal, M.N.; Raphael, K.G.; Nayak, S.; Klausner, J. Prevalence o myofascial témporo-mandibular disorder in US community women. *J. Oral Rehabil.* **2008**, *35*, 801–809. [CrossRef]
31. Mladenović, I.; Dodić, S.; Stošić, S.; Petrović, D.; Cutović, T.; Kozomara, R. TMD in class III patients referred for orthognathic surgery: Psychological and dentition-related aspects. *J. Craniomaxillofac. Surg.* **2014**, *42*, 1604–1609. [CrossRef]

 *diagnostics* 

Article

Positional Features of the Mandibular Condyle in Patients with Facial Asymmetry

Diego Fernando López [1,*], Valentina Rios Borrás [2] and Rodrigo Cárdenas-Perilla [3]

[1] Orthodontics Department, Universidad del Valle, Cali 760043, Colombia
[2] Escuela de Odontología, Universidad del Valle, Cali 760043, Colombia
[3] Nuclear Medicine Department, Clinica Imbanaco-Grupo Quirón Salud, Cali 760043, Colombia
* Correspondence: diego.f.lopez@correounivalle.edu.co

Abstract: Objective: To describe the position of the mandibular condyle, the size of the joint spaces and the condylar angulation in patients with facial asymmetry (FA), and to classify these results according to the type of FA and compare them with a reference group without FA. Materials and Methods/Patients: An observational, cross-sectional, descriptive study using computed tomography (CT) was conducted on a sample of 133 patients with a clinical diagnosis of FA derived from the following entities: hemimandibular elongation (HE) ($n = 61$), hemimandibular hyperplasia (HH) ($n = 11$), condylar hyperplasia in its hybrid form (HF) ($n = 19$), asymmetric mandibular prognathism (AMP) ($n = 25$), glenoid fossa asymmetry (GFA) ($n = 9$) and functional laterognathism (FL) ($n = 8$). Likewise, a group of 20 patients without clinical or tomographic characteristics of FA was taken and their complete cone beam tomography (CBCT) scans were analyzed. The quantified variables were joint spaces (anterior, middle and posterior), angle of the condylar axis and condylar position. All measurements were performed using the free, open-source Horos software. Results: Most of the subjects without FA had a right middle condylar position (55%), while in the patients with FA the anterior condylar position predominated. On the left side, the most frequent condylar position was anterior, including the group without FA, except in the HH group. Considering the measurements of the anterior, middle and posterior joint space (mm) on the right side (anterior JS: 1.9 mm, middle JS: 2 mm and posterior JS: 2.8 mm) and on the left side (anterior JS: 2.7 mm, middle JS: 2.1 mm and posterior JS: 2.6 mm) of the subjects without FA, compared to those with FA, the latter presented smaller distances in all diagnoses and only for the right posterior JS (1.9 mm) in HH, was not significant. The condylar axis of the AF group showed significant differences with smaller angles for the left side in those diagnosed with HE (65.4°) and HH (56.5°) compared to those without AF (70.4°). Conclusions: The condylar position of patients with FA tends to be anterior, both on the right and left sides, while for cases without FA it is middle and anterior, respectively. Patients with FA have smaller joint spaces (mm) compared to patients without FA, with the exception of HH for the right posterior JS.

Keywords: computed tomography; cone beam computed tomography; condylar anatomy; facial asymmetry; temporomandibular joint

Citation: López, D.F.; Rios Borrás, V.; Cárdenas-Perilla, R. Positional Features of the Mandibular Condyle in Patients with Facial Asymmetry. *Diagnostics* **2023**, *13*, 1034. https://doi.org/10.3390/diagnostics13061034

Academic Editor: Mara Carsote

Received: 17 February 2023
Revised: 2 March 2023
Accepted: 5 March 2023
Published: 8 March 2023

1. Introduction

The temporomandibular joint (TMJ) is a bilateral, synovial, ginglymoarthrodial structure. It has in common with other articulations of this kind the anatomic conformation with two articular surfaces (the glenoid fossa of the temporal bone and the mandibular condyle), an articular disc, the articular capsule, ligaments and synovial liquid. A distinctive characteristic of this joint is that its articular surfaces are covered by fibrocartilage [1–3]. The TMJ is a highly functional demand articulation and therefore it is susceptible to present painful symptoms of different etiologies, including muscular and capsular alterations, ligament disorders, altered masticatory patterns and changes in the position of bone components and the articular disk. Therefore, under altered anatomic conditions, functional changes causing temporomandibular disorders (TMD) and/or inner articular damage are expected [3].

Although the TMJ has a high adaptive ability, the anatomic position may be affected by pathologic entities that generate substantial morphologic changes in facial expression, as is the case of facial asymmetry (FA) related to unilateral condylar hyperplasia (UCH) [4].

UCH is due to excessive growth of a mandibular condyle generated by bone metabolic hyperactivity. It is a self-limited condition, frequently unilateral, with esthetic, occlusal and functional consequences derived from the change in mandibular position [4,5].

Another entity altering the position of the articular fossa is glenoid fossa asymmetry (GFA), which is evident during the first years of development and is a defect in the proliferation, migration and differentiation of neural crest cells [6,7].

Adaptive remodeling following a severe trauma is another possible cause of asymmetry with no alteration of the mandibular condyle [8].

The mandibular anatomy may be altered as well by mandibular asymmetric prognathism (MAP), due in this case to a bilateral difference in the effective size of the mandible. MAP etiology is genetic, and it is evident during the first stages of dental development and develops skeletal Class III [9].

Finally, functional laterognathism (FL) is an entity causing FA related to changes in mandibular position appearing early at the occlusal level and characterized as a secondary adaptation of the mandible to a disbalance in the skeletal and occluso-maxillo-mandibular relationship. This secondary adaptation, if it is not early treated negatively evolves during growth to a true skeletal asymmetry with no differences in size of the condyle skeletal components or in the mandibular ramus [10,11].

Although some authors [12–14] suggest that the inner TMJ deterioration and a severe TMD may be predisposing factors to asymmetry in mandibular position, the relationship between the pathologies generating FA and the presence or absence of TMD is not well established because some alterations are present with no evidence of articular signs or symptoms [15].

The literature reports TMD patients without FA, presenting changes in the position of the mandibular condyles, characterized by a more posterior displacement of them [16]. Regarding dimensional changes in the articular spaces, some authors associate the reduction in superior and posterior space, as well as the increment in the anterior space, to anterior displacement of the disk in patients with no significant FA [17,18]. However, condylar position and the size of articular spaces in relation to TMD is a controversial subject. In patients with asymmetry, no reports were found indicating a significant difference in the angle formed by the latero-medial plane of each condyle and the mid-sagittal plane (MSP) [19].

The most effective way to evaluate the position of all the TMJ components is through a tomographic image able to detect sagittal, coronal and axial changes [20]. Therefore, the objective of this study was to use computed tomography (CT) in a group of patients with FA and cone beam computed tomography (CBCT) in non-asymmetric subjects, to obtain linear and angular measurements of the mandibular condyle position with respect to the articular cavity.

2. Materials and Methods

There was no risk research, using only retrospective documental data with no intervention. This study was approved by the Institutional Ethics Committees involved (Clínica Imbanaco: CEI-545 and Universidad del Valle: 032-021) and it was conducted according to the principles of the Declaration of Helsinki.

The CT data (Figure 1) of 133 patients treated in a clinical center of high complexity (Imbanaco) during January 2015 and January 2020 were evaluated. The inclusion criteria were diagnosis of FA and complete and acceptable CT images. The exclusion criteria were antecedents of TMJ pathology and/or surgery, trauma or fracture, treatment with occlusal splints, orthognathic surgery, dentofacial syndromic anomalies, arthritis and incomplete CT studies.

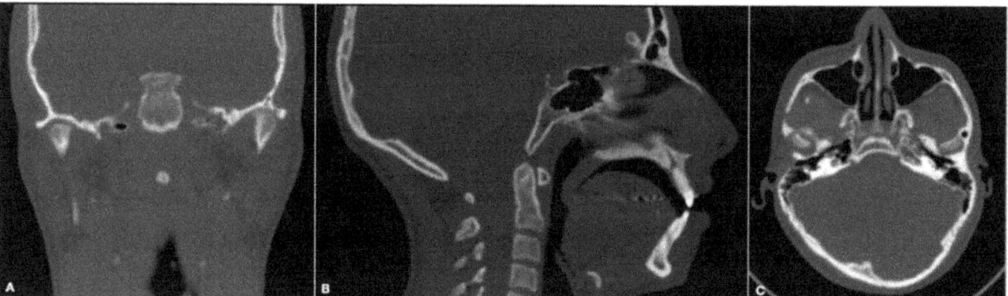

Figure 1. CT of a patient with left side condylar hyperplasia. (**A**) Coronal view. (**B**) Sagittal view. (**C**) Axial view.

For UCH cases, the affected side was defined as the side with condylar overdevelopment; for MAP and FL, it was the side of mandibular deviation and for GFA it was the side with evident upper projection of the articular cavity.

CT images were obtained with PET/CT Biograph mCT20 (Siemens, Erlangen, Germany) equipment. Cranial images were obtained without contrast media, from vertex to sternal fork, applying the following parameters: section thickness 0.75 mm, pitch 1.0 and cubic matrix 512 × 512, isotropic voxel (size: 0.58 × 0.58 × 0.87 mm) to avoid image distortion in adult and growing patients. CT images were reconstructed using a B26F homogeneous, low-dose filter for anatomic location. All the patients were positioned with fixed head to avoid movement artifacts and facilitate image fusion.

The CBCT images (Figure 2) of 20 patients scheduled to initiate orthodontic treatment, with no mandibular deviation, suspected FA or TMD signs, obtained from April 2019 to March 2022, were selected from the Oral Radiology Department of the Universidad del Valle, Cali, Colombia.

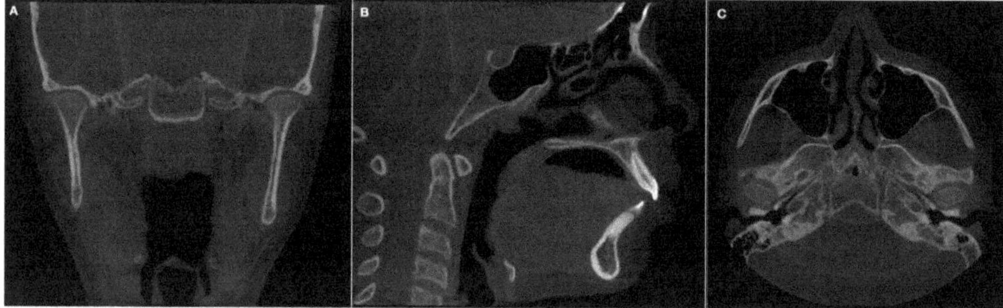

Figure 2. CBCT from a patient with no asymmetry (non-asymmetric group). (**A**) Coronal view. (**B**) Sagittal view. (**C**) Axial view.

CBCT images were obtained with i-CAT 17-19 equipment. Cranial images were obtained with no use of contrast media, from nasion to menton. The patients were in corrected natural head position. The following parameters were applied: camp window (FOV): 16 cm, width 0.250 mm, isotropic voxel (size: 0.25 × 0.25 × 0.25 mm) to avoid any image distortion in adult and growing patients.

The CT and CBCT images were stored in digital form and digital communications in medicine (DICOM). The DICOM 2D images were downloaded to the Horos software for processing, visualization and bidimensional measurement of the anatomic structures as described in Table 1.

Table 1. Description of the variables measured in CT and CBCT.

Variable		Description
Articular space	Posterior	Draw a tangent line to the posterior wall of the mandibular condyle. The most posterior-superior point of the condyle on the tangent line is located and from that point a perpendicular is traced to the posterior wall of the articular cavity. Data in mm. Figure 3c.
	Middle	Draw a line from the uppermost point of the mandibular condyle to the deepest point of the glenoid fossa. Data in mm. Figure 3d.
	Anterior	Draw a tangent line from the most anterior part of the mandibular condyle. The most antero-superior point of the condyle on the tangent line is located and from that point a perpendicular is traced to the anterior wall of the articular cavity. Data in mm. Figure 3e.
Condylar Position		Apply the equation: DC = (P − A/P + A) × 100%. DC (condylar displacement). P (posterior articular space). A (Anterior articular space). Method described by Pullinger and Hollender, modified by Pereira et al., 2007 [21,22].
Condylar axis		Inner angle between sagittal middle plane (SMP) and the line drawn from a projection of the highest middle-lateral length of each mandibular condyle. Figure 4.

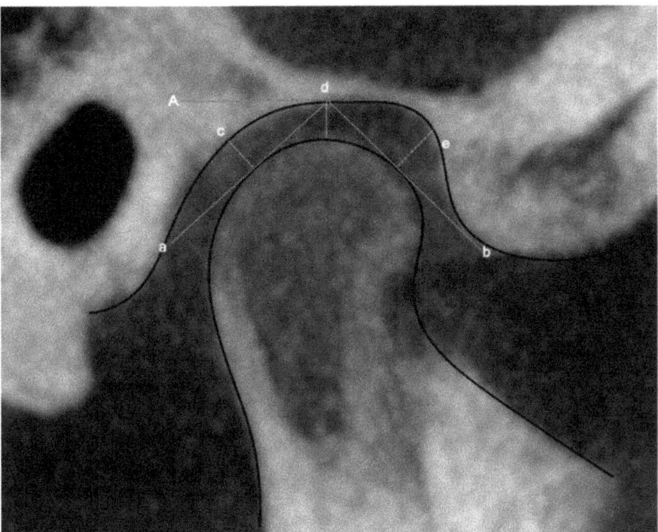

Figure 3. References for linear measurement of articular spaces (modified from Ikeda and Kawamura, 2009) [24]. **A.** True horizontal crossing the deepest point of glenoid fossa. **a.** Tangent line crossing the most posterior point of the mandibular condyle and the deepest point of the glenoid fossa. **b.** Tangent line crossing the most anterior point of the mandibular condyle and the deepest point of the glenoid fossa. **c.** Perpendicular from the most postero-superior point of the condyle on the tangent to the posterior wall of the glenoid fossa. **d.** Line from the most superior point of the mandibular condyle to the deepest point of the glenoid fossa. **e.** Perpendicular from the most antero-superior point of the condyle on the tangent to the anterior wall of the glenoid fossa. The distance in mm in these lines measures the anterior, middle and posterior articular space.

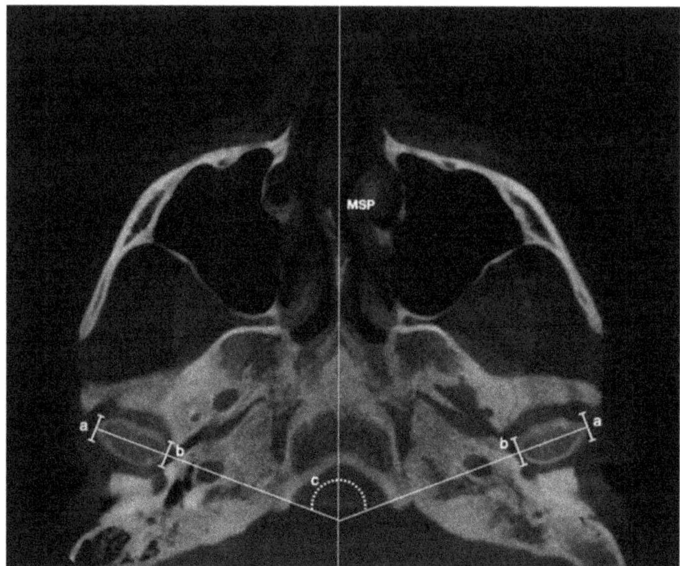

Figure 4. References to obtain the angular measurement of condylar axis with respect to the mid-sagittal plane (MSP). **a.** Lateral extreme of the mandibular condyle. **b.** Medial extreme of the mandibular condyle. **c.** Inner angle between MSP and the line drawn from the projection of higher mid-lateral length of each condyle.

The measurements were registered by an operator expert in the software management and TMJ anatomy. Each set of images and data was evaluated and classified under operator and clinician agreement, according to the craniofacial characteristics of the asymmetry [23], (Table 2).

Table 2. Intraobserver agreement for patients treated during the period 2015–2020 with CT indicated due to FA.

Variable	Measurement 1 ($n = 20$) *	Measurement 2 ($n = 20$) *	CCA **
Right joint space			
Anterior	1 (0.7; 1.2)	1 (0.7; 1.3)	0.89
Middle	1.2 (0.9; 1.9)	1.3 (0.7; 1.7)	0.93
Posterior	1.4 (1.3; 1.6)	1.5 (1.2; 1.7)	0.96
Left joint space			
Anterior	1.1 (0.9; 1.5)	1.3 (0.9; 1.6)	0.92
Middle	1.3 (1; 1.8)	1.2 (0.9; 1.7)	0.93
Posterior	1.4 (1.1; 1.8)	1.6 (1.2; 2)	0.92
Condylar axis			
Right	66.4 (60.5; 72.2)	66.7 (60.5; 72.4)	1
Left	65.4 (63.1; 74.1)	65.7 (63.8; 73.3)	1

* Median (p25; p75), ** Correlation coefficient of agreement.

Statistical Analysis

Descriptive data were presented as central tendency (mean, median) and standard deviation or P25-P75, following the Shapiro–Wilk normality test for parametric variables. Non-parametric variables are expressed as absolute and relative percentage frequency.

Initially the intraoperator agreement was estimated by the correlation coefficient of agreement (CCA), obtaining a CCA value of 89% for the right anterior space and >90% for the other data and the condylar axis angle. Comparative tests (chi-square, *t*-test or U test) were applied as necessary. Any p value < 0.05 was accepted as significant. The statistical program used was R 4.2.2.

3. Results

Considering the selection criteria, a database of 133 patients with a diagnosis of FA was obtained. Additionally, the data of 20 non-asymmetric orthodontic patients with no FA or signs/symptoms of TMD were included. The median age for the non-asymmetric group of subjects was 22 years, and for the FA group age was in a range of 14–26, with the lower median (14 years) in the GFA group and the highest median (26 years) in the HH group. However, 75% of the patients were under 30 years. Female gender represented 61.4% of the total sample. The right side was more frequently affected (51.4%) and the more frequent diagnosis was HE, representing 45.9% of the asymmetry group. (Table 3).

Table 3. Demographic and clinical description of the patients with FA and subjects without FA or signs of TMD.

Variable (n = 153)	NA (n = 20)	MAP (n = 25)	GFA (n = 9)	FL (n = 8)	HE (n = 61)	HH (n = 11)	HF (n = 19)
Age *	22 (16; 29)	17 (13; 19)	14 (13; 16)	16 (14; 20)	17 (15; 24)	26 (17; 30)	23 (16; 30)
Gender **							
Male	8 (40)	12 (48)	2 (22.2)	3 (37.5)	24 (39.3)	2 (18.2)	8 (42.1)
Female	12 (60)	13 (52)	7 (77.8)	5 (62.5)	37 (60.7)	9 (81.8)	11 (57.9)
Affected side **							
Right	0 (0)	15 (60)	4 (44.4)	3 (37.5)	32 (52.5)	7 (63.6)	11 (57.9)
Left	0 (0)	10 (40)	5 (55.6)	5 (62.5)	29 (47.5)	4 (36.4)	8 (42.1)

NA: non-asymmetric, MAP: mandibular asymmetric prognathism, GFA: glenoid fossa asymmetry, FL: functional laterognathism, HE: hemimandibular elongation, HH: hemimandibular hyperplasia, HF: hybrid form. * Median (p25; p75), ** n (%).

According to the classification published by López et al. [23], the kind of FA was established as: condylar hyperplasia (CH): 91 cases (61 HE, 11 HH and 19 HF), MAP: 29, GFA: 9 and FL: 8.

In Table 4, the sample is regrouped according to the condylar position: posterior: <12%, middle: −12 to 12% and anterior >12%.

Table 4. Condylar position in FA patients and non-asymmetric subjects without FA.

Condylar Position (%) (n = 153)	NA (n = 20)	MAP (n = 25)	GFA (n = 9)	FL (n = 8)	HE (n = 61)	HH (n = 11)	HF (n = 19)
Right side							
Anterior	7 (35)	16 (64)	7 (77.8)	6 (75)	27 (44.3)	10 (90.9)	12 (63.2)
Middle	11 (55)	8 (32)	2 (22.2)	2 (25)	19 (31.1)	0 (0)	5 (26.3)
Posterior	2 (10)	1 (4)	0 (0)	0 (0)	15 (24.6)	1 (9.1)	2 (10.5)
Left side							
Anterior	10 (50)	13 (52)	6 (66.7)	5 (62.5)	33 (54.1)	4 (36.4)	7 (36.8)
Middle	8 (40)	6 (24)	3 (33.3)	1 (12.5)	18 (29.5)	5 (45.5)	7 (36.8)
Posterior	2 (10)	6 (24)	0 (0)	2 (25)	10 (16.4)	2 (18.2)	5 (26.3)

NA: non-asymmetric, MAP: mandibular asymmetric prognathism, GFA: glenoid fossa asymmetry, FL: functional laterognathism, HE: hemimandibular elongation, HH: hemimandibular hyperplasia, HF: hybrid form.

It was found that for the right-side data most non-asymmetric subjects had middle position (55%), while in patients with FA diagnosis the anterior condylar position was the most frequent, between 44 to 91% depending on the kind of FA. On the left side for both groups (FA and no FA), the anterior condylar position was the most frequent, except in the HH group.

Regarding the condylar position (%) in both sides, when the FA patients were compared to the without FA group (right: 7.4% and left: 11.7%) significant differences were found in the MAP group (15.7%). However, when comparing only the affected side, the differences were not significant in MAP patients. In the HH group, compared to the without FA group, the difference is significant for the right side (33.3%, $p < 0.05$) and in general ($p < 0.01$). In the GFA group, there was significant difference in the left side (30.8%, $p < 0.05$). (Table 5).

Table 5. Comparison of condylar position between affected sides and kind of FA vs. without FA.

Diagnosis and Affected Side (n = 153)	Right Condylar Position (%)	p Value	Left Condylar Position (%)	p Value
NA	7.4 (−3; 17) *	Ref	11.7 (0.3; 24.1) *	Ref
MAP	15.7 (7; 33.3) *	0.04	12.5 (−4.3; 36.3) *	0.85
Right	29.5 (0.6; 35.8) *	0.16	-	-
Left	-	-	15 (−13.1; 33.2) *	1
GFA	25.5 (22.6; 28.4) *	0.06	24.3 (9.8; 50.4) *	0.13
Right	20.1 (-) **	0.24	-	-
Left	-	-	30.8 (24.3; 54.5) *	0.03
FL	19.9 (13.2; 29.8) *	0.08	24.3 (−2.6; 33.6) *	0.57
Right	22.8 (-) **	0.12	-	-
Left	-	-	27.5 (-) **	0.15
HE	8.7 (−9.9; 23) *	0.77	15.3 (−5.5; 30.5) *	0.93
Right	13.7 (0.6; 20.4) *	0.46	-	-
Left	-	-	15.3 (1.5; 30.6) *	0.64
HH	33.3 (25.3; 42.3) *	<0.01	7.1 (−6.8; 30.4) *	0.64
Right	33.3 (22.3; 40.4) *	0.03	-	-
Left	-	-	27.2 (-) **	0.48
HF	23.9 (2.3; 38.2) *	0.08	−1.7 (−12.4; 25.4) *	0.29
Right	18.2 (2.3; 36.4) *	0.23	-	-
Left	-	-	3.5 (−14.2; 15.6) *	0.3

* Median (p25; p75), ** Average. (For sample sizes < 5 Q range was not calculated.) NA: non-asymmetric, MAP: mandibular asymmetric prognathism, GFA: glenoid fossa asymmetry, FL: functional laterognathism, HE: hemimandibular elongation, HH: hemimandibular hyperplasia, HF: hybrid form. n (%).

Table 6 shows that, comparing the measurements of the condylar axis (°) of non-asymmetric subjects (right: 68.9° and left: 70.4°) versus patients with FA, there are only significant differences in measurements on the left side of those diagnosed with EH (65.4°) and HH (56.5°), without disaggregating by affected side.

Taking into account the measurements of the anterior, middle and posterior joint space (mm) on the right side, it is observed that when comparing the non-asymmetric subjects (anterior JS: 1.9 mm, middle JS: 2 mm and posterior JS: 2.8 mm) with the measurements of patients with FA, the latter present smaller distances with statistically significant differences in all diagnoses and only for the posterior joint space (1.9 mm) in HH, it is not significant. Additionally, when they are analyzed by the affected right side, statistically significant differences are found in most entities, with the exception of the GFA, which has a sample size of only 4 cases.

For the measurements of the left side, it is observed that when comparing the non-asymmetric subjects (anterior JS: 2.7 mm, middle JS: 2.1 mm and posterior JS: 2.6 mm), with those diagnosed with asymmetry, there are also significant differences in all entities, presenting the latter shorter distances. When disaggregating with respect to the left affected side, all those diagnosed with FA have minor joint spaces with statistically significant differences and only the posterior joint space of the GFA (2.1 mm), FL (2.1 mm) and HH (2 mm), were not significant, however, there are samples of less than 5 in these cases (Table 7).

Table 6. Comparison of condylar axis angle data by sides in without FA group vs. FA groups.

Diagnosis and Affected Side ($n = 153$)	Right Condylar Axis (°)	*p* Value	Left Condylar Axis (°)	*p* Value
NA	68.9 (61.3; 73.3) *	Ref	70.4 (64.7; 75) *	Ref
MAP	69.5 (63.8; 74.5) *	0.5	69.4 (64.9; 72.2) *	0.78
Right	70.7 (65.2; 76.3) *	0.27	-	-
Left	-	-	69.1 (67.1; 71.3) *	0.91
GFA	66 (62.9; 74.7) *	0.94	66.4 (63.5; 73) *	0.44
Right	63.7 (-) **	0.31	-	-
Left	-	-	67.7 (-) **	0.62
FL	63.6 (59.2; 72) *	0.5	66.9 (61; 73.7) *	0.6
Right	66.4 (-) **	0.9	-	-
Left	-	-	63.1 (-) **	0.57
HE	65.2 (59.5; 72.9) *	0.41	65.4 (61.8; 70.3) *	0.03
Right	69.1 (61.4; 75) *	0.57	-	-
Left	-	-	65.7 (62.7; 71.7) *	0.14
HH	60.6 (51.7; 65.5) *	0.06	56.5 (51.4; 66.1) *	<0.01
Right	61.3 (60.2; 69.5) *	0.53	-	-
Left	-	-	63.6 (-) **	0.27
HF	62 (55.1; 69.9) *	0.21	62.2 (55.6; 74.2) *	0.11
Right	65.7 (62; 73.8) *	0.92	-	-
Left	-	-	64 (60.2; 77.6) *	0.64

NA: non-asymmetric * Median (p25; p75), ** Average. (For sample sizes < 5 Q range was not calculated). MAP: mandibular asymmetric prognathism, GFA: glenoid fossa asymmetry, FL: functional laterognathism, HE: hemimandibular elongation, HH: hemimandibular hyperplasia, HF: hybrid form. *n* (%).

Table 7. Comparison of joint space data of without FA group vs. FA groups and sides.

Diagnosis and Affected Side ($n = 153$)	Right Anterior (mm)	*p*	Right Middle (mm)	*p*	Right Posterior (mm)	*p*	Left Anterior (mm)	*p*	Left Middle (mm)	*p*	Left Posterior (mm)	*p*
NA	1.9 (1.6; 2.3) *	Ref	2 (1.7; 2.3) *	Ref	2.8 (2.2; 3.6) *	Ref	2.7 (2.3; 3) *	Ref	2.1 (1.8; 2.8) *	Ref	2.6 (2; 3) *	Ref
MAP	0.9 (0.6; 1.3) *	<0.01	1.1 (0.8; 1.9) *	<0.01	1.5 (1.1; 1.8) *	<0.01	1.2 (0.9; 1.4) *	<0.01	1.5 (0.9; 1.8) *	<0.01	1.5 (1.2; 1.8) *	<0.01
Right	0.9 (0.6; 1.1) *	<0.01	1.1 (0.8; 1.9) *	<0.01	1.2 (1; 1.8) *	<0.01	-	-	-	-	-	-
Left	-	-	-	-	-	-	1.2 (1.1; 1.4) *	0.01	1.7 (1.5; 2) *	<0.01	1.6 (1.4; 2) *	<0.01
GFA	0.9 (0.8; 0.9) *	<0.01	1.3 (0.9; 1.5) *	<0.01	1.4 (1.3; 1.6) *	<0.01	1 (0.8; 1)	<0.01	1.6 (1.2; 2.3)	0.02	1.5 (1.4; 1.9) *	0.03
Right	1.2 (-) *	0.06	2 (-) **	0.13	1.8 (-) **	0.1	-	-	-	-	-	-
Left	-	-	-	-	-	-	0.8 (-) **	<0.01	1.5 (-) **	0.01	2.1 (-) **	0.1
FL	1 (0.7; 1.3) *	<0.01	1.3 (1.2; 1.6) *	<0.01	1.6 (1.4; 1.7) *	<0.01	1.2 (0.9; 1.8)	0.04	1.5 (1.1; 1.8) *	<0.01	1.9 (1.2; 2.2) *	0.02
Right	0.9 (-) **	0.01	1.2 (-) **	0.01	1.5 (-) **	<0.01	-	-	-	-	-	-
Left	-	-	-	-	-	-	1 (-) **	0.02	1.6 (-) **	0.01	2.1 (-) **	0.3
HE	1.1 (0.9; 1.4) *	<0.01	1.4 (0.9; 1.8) *	<0.01	1.3 (1.1; 1.7) *	<0.01	1.1 (0.9; 1.5) *	<0.01	1.4 (0.9; 1.7) *	<0.01	1.4 (1.2; 1.8) *	<0.01
Right	1 (0.9; 1.3) *	<0.01	1.2 (0.9; 1.7) *	<0.01	1.3 (1.1; 1.5) *	<0.01	-	-	-	-	-	-
Left	-	-	-	-	-	-	1.1 (0.9; 1.2) *	<0.01	1.1 (0.8; 1.5) *	<0.01	1.6 (1.2; 1.7) *	<0.01
HH	0.9 (0.8; 1.0) *	<0.01	0.8 (0.7; 2.0) *	<0.01	1.9 (1.3; 2.2) *	0.16	1.5 (1.1; 1.8) *	<0.01	1.2 (0.8; 1.5) *	<0.01	1.7 (1.4; 1.9) *	<0.01
Right	1 (0.9; 1.1) *	<0.01	1.2 (0.7; 2.4) *	0.01	2 (1.6; 2.2) *	0.26	-	-	-	-	-	-
Left	-	-	-	-	-	-	1.1 (-) **	0.03	0.9 (-) **	<0.01	2 (-) **	0.08
HF	1 (0.7; 1.2) *	<0.01	1 (0.8; 1.6) *	<0.01	1.5 (1.3; 1.7) *	<0.01	1.2 (0.9; 1.7) *	<0.01	1.3 (0.9; 1.7) *	<0.01	1.4 (1.1; 1.6) *	<0.01
Right	1 (0.8; 1.2) *	<0.01	1 (0.9; 1.6) *	<0.01	1.4 (1.3; 1.5) *	<0.01	-	-	-	-	-	-
Left	-	-	-	-	-	-	1.1 (1; 1.5) *	<0.01	1.1 (0.7; 1.6) *	<0.01	1.2 (1; 1.3) *	<0.01

NA: non-asymmetric, MAP: mandibular asymmetric prognathism, GFA: glenoid fossa asymmetry, FL: functional laterognathism, HE: hemimandibular elongation, HH: hemimandibular hyperplasia, HF: hybrid form. *n* (%), * Median (p25; p75), ** Average. (For sample sizes < 5 Q range was not calculated).

4. Discussion

The spatial orientation of the mandibular condyle with respect to the joint cavity in the TMJ may be influenced by anatomical, functional and/or pathological characteristics [25]. Some studies have evaluated these characteristics and their relationship with joint disorders or TMD, but few have focused on patients with structural skeletal disorders such as patients with FA. In the present study, the characteristics related to the condyle and its articular cavity in patients with different entities causing FA were evaluated.

With respect to joint spaces, Ikeda et al. [24] determined in their study with CBCT mean values in non-asymmetric patients, where the anterior, middle and posterior JS were 1,3, 2,5 and 2,1 mm, respectively. These values are close to those found in non-asymmetric patients in the present study, and far from those found in asymmetric patients. In this study, it was evidenced that the population with FA, regardless of the entity that produces the alteration, has smaller joint spaces than those patients without FA. Regarding this, Major et al. [25] reported that alterations in the joint spaces were associated with anterior displacement of the disc and a decrease in its length, although in their study of growing patients, the decrease in joint space was limited to the medial space. Likewise, A. K. Bag et al. [26] reported the possible association between the decrease in joint spaces with unilateral and bilateral disc displacements.

This means that if the function of the articular disc, in addition to supporting joint loads, is to provide synovial fluid to the bone surfaces that helps its nutrition, oxygenation, lubrication and hydration [27], the possible displacement and alteration in its anatomy would mean equally pathological and functional changes [28].

In fact, it has been hypothesized that the reduction in the joint space affects the condylar position in the contralateral TMJ [29]. However, the assessment of the joint space by itself is not enough to determine whether or not there is presence of TMD [30].

Likewise, in the present study, when the values are analyzed by affected side and by each entity, with the exception of GFA, all joint spaces were smaller. It was even evident in cases of condylar hyperplasia, in which there is a substantial change in the condylar size and the height of the joint cavity towards the affected side [31]. It was evidenced that the joint spaces were decreased with respect to the non-asymmetric subjects and only for the posterior joint space it was not significant in HH. The lack of significance may be explained by the small number of cases (n = 4).

Regarding the demographic characteristics, the majority of patients with FA were women and the most affected side was the right side. This is coincident with prior studies published by Raijmakers et al. [32] and López et al. [33].

The condylar position showed a higher percentage of middle condylar position in the right condyle for non-asymmetric subjects, while for asymmetric patients it was predominantly anterior, independent of the kind of FA. An interesting observation was that the right side was affected in 54.1% of the patients. On the other hand, when the left side was analyzed, all presented an anterior condylar position, including the non-asymmetric subjects, and it was not evident only for the four cases of HH. Similar results with differences between sides were obtained by Chae et al. [34] in an adolescent population and with a predominance of anterior condylar position in the left joint in the study of Ganugapanta et al. [35].

The comparison of each entity of FA and the without FA subjects was significant in MAP and HH groups only for the right side. The position in these cases was anterior as well. The lack of coincidence between sides and between with or without FA groups is coincident with the reports published by Paknahad et al. [36] and Guerrero et al. [37], showing that there are no differences in the condylar position in patients with or without TMD. Additionally, Lelis et al. [38] did not find differences between symptomatic and asymptomatic patients; as reported by Choi et al. [39], even in patients who underwent orthognathic surgery from sagittal mandibular osteotomies to correct mandibular prognathism and facial asymmetry, no changes in the condylar position were observed after surgery.

Differences have been reported for specific malocclusions such as the anterior open bite and posterior cross-bite, which show posterior condylar positions. [40]. Skeletal discrepancies in Class II subjects, show antero-superior condylar positions and hyperdivergent patterns with higher risk of condylar displacement [41,42].

In patients with asymmetry and a resulting posterior crossbite, as is the case with HE and FH, and which can also occur in cases of PMA, LF and even in GFA, Almaqrami et al. [43] postulate that skeletal crossbite is accompanied by morphologic and positional features in the TMJ associated with dental unilateral posterior crossbite and are associated with specific asymmetry on one side of the TMJ. In the present study differences between sides were not significant, but the condylar position was measured only in the sagittal plane, not in the transaxial.

In relation to the condylar axis, Westesson et al. [44] described a more closed axial condylar angle in normal TMJs, while for affected joints, such as those with disc displacement, this angle was much more open. Regarding this, Al Rawi et al. [30], found that there were differences between men and women for the angle of the condylar axis, being more closed in women. Unlike the findings of Westesson et al. [44], however, the angle of the condyle axis tended to decrease significantly in patients with TMD, both for men and women, showing internal rotation of the condyle in affected TMJs.

In the present study, differences were only found with respect to the non-asymmetric subjects for HE and HH in the left condylar axis in general, presenting smaller angles, but when disaggregated by the affected side, no differences were found. In this regard, it is worth mentioning that the universe of the present sample was patients with FA and not TMD. Similar results are reported by Rodrigues et al. [19] evaluating the angle between the latero-medial plane of each condyle and the mid-sagittal plane in Class I patients with no FA. The bilateral comparison of this angle shows mean values very similar to those of the non-asymmetric subjects in the present study (right side 70.10° and left side 69.96°).

Although it is well accepted that CT and CBCT imaging are gold standards for assessing morphologic and structural features of craniofacial bones and TMJ [45,46], they lack sensitivity for assessing soft tissues that are relevant for describing TMD. [47]. Therefore, it is suggested that future studies include joint symptoms and correlate AF with TMD. One limitation of this research is that the slice thickness of medical tomographies (0.75 mm) is an unmodifiable characteristic of the medical center since they are standardized both for accuracy and for radiation dose to this measure and cannot be with smaller slice thicknesses that, although they give more image sharpness, increase radiation.

5. Conclusions

There are marked differences between the sides in condylar position, both in patients with FA and without FA. The right side tends to have a middle position in non-asymmetric subjects and an anterior position in all FA patients. The left side has a predominantly anterior position in both AF and non-AF cases.

A greater anterior condylar position was evidenced for the right side in MAP ($p = 0.04$); for the affected left side in GFA ($p = 0.03$); and both general and for the affected right side in HH ($p < 0.01$ and $p = 0.03$), respectively, compared to the group without AF.

Patients with FA have reduced anterior, middle and posterior joint spaces with respect to non-asymmetric patients for both the right and left joints. There were no significant differences only in the right posterior joint space of the HH.

The angle of the condylar axis only showed differences for the HE and HH on the left side, these being smaller with respect to the non-asymmetric ones.

Author Contributions: Conceptualization, D.F.L.; formal analysis, R.C.-P.; investigation, D.F.L. and V.R.B.; methodology, R.C.-P. and V.R.B.; supervision, D.F.L. and R.C.-P.; validation, R.C.-P.; visualization, V.R.B.; writing—original draft, V.R.B. and R.C.-P.; writing—review and editing, D.F.L. All authors have read and agreed to the published version of the manuscript.

Funding: This research received no external funding.

Institutional Review Board Statement: This is a study with no intervention or manipulation of variables from the patients. Therefore, it is a no risk investigation and was approved by the Institutional Ethics Committee (Approval number Imbanaco Medical Center: CEI-545 and Universidad del Valle: 032-021) and conducted following all the regulations of the Declaration of Helsinki, last version.

Informed Consent Statement: Patient consent not due; this is a study with no intervention or manipulation of variables from the patients, therefore, it is a no risk investigation.

Data Availability Statement: Data available on request due to restrictions, e.g., privacy or ethical.

Acknowledgments: The authors wish to express their sincere gratitude to the Imbanaco Medical Center Research Institute staff for their support during the development of this study.

Conflicts of Interest: The authors declare no conflict of interest.

References

1. Alomar, X.; Medrano, J.; Cabratosa, J.; Clavero, J.; Lorente, M.; Serra, I.; Monill, J.; Salvador, A. Anatomy of the temporomandibular joint. *Semin. Ultrasound CT MRI* **2007**, *28*, 170–183. [CrossRef] [PubMed]
2. Wang, X.; Yang, C.; Goddard, G.; Qiu, W. Normal and pathological anatomy of the TMJ viewed by computerized panoramic arthroscopic images. *Cranio* **2003**, *21*, 196–201. [CrossRef] [PubMed]
3. Roberts, W.E.; Stocum, D.L. Part II: Temporomandibular Joint (TMJ)-Regeneration, Degeneration, and Adaptation. *Curr. Osteoporos. Rep.* **2018**, *16*, 369–379. [CrossRef] [PubMed]
4. Wolford, L.M.; Movahed, R.; Perez, D.E. A classification system for conditions causing condylar hyperplasia. *J. Oral Maxillofac. Surg.* **2014**, *72*, 567–595. [CrossRef]
5. Kim, J.-Y.; Kim, B.-J.; Park, K.-H.; Huh, J.-K. Comparison of volume and position of the temporomandibular joint structures in patients with mandibular asymmetry. *Oral Surg. Oral Med. Oral Pathol. Oral Radiol.* **2016**, *122*, 772–780. [CrossRef]
6. López Buitrago, D.F.; Ruiz Botero, J. Asymmetry of glenoid fossa as differential diagnosis for hemimandibular elongation. *Rev. Mex. Ortod.* **2017**, *5*, e217–e226. [CrossRef]
7. Minoux, M.; Rijli, F.M. Molecular mechanisms of cranial neural crest cell migration and patterning in craniofacial development. *Development* **2010**, *137*, 2605–2621. [CrossRef]
8. Baek, S.-H.; Cho, I.-S.; Chang, Y.-I.; Kim, M.-J. Skeletodental factors affecting chin point deviation in female patients with class III malocclusion and facial asymmetry: A three-dimensional analysis using computed tomography. *Oral Surg. Oral Med. Oral Pathol. Oral Radiol. Endodontol.* **2007**, *104*, 628–639. [CrossRef]
9. Pirttiniemi, P.M. Associations of mandibular and facial asymmetries—A review. *Am. J. Orthod. Dentofac. Orthop.* **1994**, *106*, 191–200. [CrossRef]
10. Ishizaki, K.; Suzuki, K.; Mito, T.; Tanaka, E.M.; Sato, S. Morphologic, functional, and occlusal characterization of mandibular lateral displacement malocclusion. *Am. J. Orthod. Dentofac. Orthop.* **2010**, *137*, 454.e1–454.e9. [CrossRef]
11. Schmid, W.; Mongini, F.; Felisio, A. A computer-based assessment of structural and displacement asymmetries of the mandible. *Am. J. Orthod. Dentofac. Orthop.* **1991**, *100*, 19–34. [CrossRef] [PubMed]
12. Katzberg, R.W.; Tallents, R.H.; Hayakawa, K.; Miller, T.L.; Goske, M.J.; Wood, B.P. Internal derangements of the temporomandibular joint: Findings in the pediatric age group. *Radiology* **1985**, *154*, 125–127. [CrossRef] [PubMed]
13. Tallents, R.H.; Guay, J.A.; Katzberg, R.W.; Murphy, W.; Proskin, H. Angular and linear comparisons with unilateral mandibular asymmetry. *J. Craniomandib. Disord. Facial Oral Pain* **1991**, *5*, 135–142.
14. Nickerson, J.W.J.; Møystad, A. Observations on individuals with radiographic bilateral condylar remodeling. *J. Craniomandib. Pract.* **1982**, *1*, 20–37. [CrossRef] [PubMed]
15. Ahn, S.-J.; Lee, S.-P.; Nahm, D.-S. Relationship between temporomandibular joint internal derangement and facial asymmetry in women. *Am. J. Orthod. Dentofac. Orthop.* **2005**, *128*, 583–591. [CrossRef]
16. Pereira, L.J.; Gavião, M.B.D. Tomographic evaluation of TMJ in adolescents with temporomandibular disorders. *Braz. Oral Res.* **2004**, *18*, 208–214. [CrossRef]
17. Matos, J.L.; Matos, M.; Ramos-Perez, F.; De Almeida, S.; Bóscolo, F. Avaliação do espaço articular em repouso de pacientes sintomáticos e assintomáticos de desordens temporomandibulares. *Braz. Dent. Sci.* **2010**, *8*, 68–76. [CrossRef]
18. Kinniburgh, R.D.; Major, P.W.; Nebbe, B.; West, K.; Glover, K.E. Osseous morphology and spatial relationships of the temporomandibular joint: Comparisons of normal and anterior disc positions. *Angle Orthod.* **2000**, *70*, 70–80.
19. Rodrigues, A.F.; Fraga, M.R.; Vitral, R.W.F. Computed tomography evaluation of the temporomandibular joint in Class I malocclusion patients: Condylar symmetry and condyle-fossa relationship. *Am. J. Orthod. Dentofac. Orthop.* **2009**, *136*, 192–198. [CrossRef]
20. Ikeda, K.; Kawamura, A.; Ikeda, R. Assessment of Optimal Condylar Position in the Coronal and Axial Planes with Limited Cone-Beam Computed Tomography. *J. Prosthodont.* **2011**, *20*, 432–438. [CrossRef]
21. Pullinger, A.G.; Hollender, L.; Solberg, W.K.; Petersson, A. A tomographic study of mandibular condyle position in an asymptomatic population. *J. Prosthet. Dent.* **1985**, *53*, 706–713. [CrossRef] [PubMed]
22. Pereira, L.J.; Gavião, M.; Bonjardim, L.; Castelo, P.M. Ultrasound and tomographic evaluation of temporomandibular joints in adolescents with and without signs and symptoms of temporomandibular disorders: A pilot study. *Dentomaxillofac. Radiol.* **2007**, *36*, 402–408. [CrossRef] [PubMed]
23. López, D.F.; Botero, J.R.; Muñoz, J.M.; Cárdenas-Perilla, R.; Moreno, M. Are There Mandibular Morphological Differences in the Various Facial Asymmetry Etiologies? A Tomographic Three-Dimensional Reconstruction Study. *J. Oral Maxillofac. Surg.* **2019**, *77*, 2324–2338. [CrossRef] [PubMed]
24. Ikeda, K.; Kawamura, A. Assessment of optimal condylar position with limited cone-beam computed tomography. *Am. J. Orthod. Dentofac. Orthop.* **2009**, *135*, 495–501. [CrossRef] [PubMed]

25. Major, P.W.; Kinniburgh, R.D.; Nebbe, B.; Prasad, N.G.; Glover, K.E. Tomographic assessment of temporomandibular joint osseous articular surface contour and spatial relationships associated with disc displacement and disc length. *Am. J. Orthod. Dentofac. Orthop.* **2002**, *121*, 152–161. [CrossRef]
26. Bag, A.K.; Gaddikeri, S.; Singhal, A.; Hardin, S.; Tran, B.D.; Medina, J.A.; Curé, J.K. Imaging of the temporomandibular joint: An update. *World J. Radiol.* **2014**, *6*, 567–582. [CrossRef]
27. Luder, H.U. Articular degeneration and remodeling in human temporomandibular joints with normal and abnormal disc position. *J. Orofac. Pain* **1993**, *7*, 391–402.
28. Panchbhai, A. Temporomandibular joint space. *Indian J. Oral Health Res* **2017**, *3*, 47–56. [CrossRef]
29. Rammelsberg, P.; Jäger, L.; Duc, J.-M.P. Magnetic resonance imaging-based joint space measurements in temporomandibular joints with disk displacements and in controls. *Oral Surg. Oral Med. Oral Pathol. Oral Radiol. Endodontol.* **2000**, *90*, 240–248. [CrossRef]
30. Al-Rawi, N.H.; Uthman, A.T.; Sodeify, S.M. Spatial analysis of mandibular condyles in patients with temporomandibular disorders and normal controls using cone beam computed tomography. *Eur. J. Dent.* **2017**, *11*, 099–105. [CrossRef]
31. López, D.F.; Botero, J.R.; Muñoz, J.M.; Cardenas-Perilla, R.A. Mandibular and temporomandibular morphologic characteristics of patients with suspected unilateral condylar hyperplasia: A CT study. *Dent. Press J. Orthod.* **2020**, *25*, 61–68. [CrossRef] [PubMed]
32. Raijmakers, P.G.; Karssemakers, L.H.; Tuinzing, D.B. Female Predominance and Effect of Gender on Unilateral Condylar Hyperplasia: A Review and Meta-Analysis. *J. Oral Maxillofac. Surg.* **2012**, *70*, e72–e76. [CrossRef]
33. Lópezb, D.F.; Corrals, C.M. Comparison of planar bone scintigraphy and single photon emission computed tomography for diagnosis of active condylar hyperplasia. *J. Cranio-Maxillofac. Surg.* **2016**, *44*, 70–74. [CrossRef] [PubMed]
34. Chae, J.-M.; Park, J.H.; Tai, K.; Mizutani, K.; Uzuka, S.; Miyashita, W.; Seo, H.Y. Evaluation of condyle-fossa relationships in adolescents with various skeletal patterns using cone-beam computed tomography. *Angle Orthod.* **2020**, *90*, 224–232. [CrossRef] [PubMed]
35. Ganugapanta, V.R.; Ponnada, S.R.; Gaddam, K.P.R.; Perumalla, K.; Khan, I.; Mohammed, N.A. Computed Tomographic Evaluation of Condylar Symmetry and Condyle-Fossa Relationship of the Temporomandibular Joint in Subjects with Normal Occlusion and Malocclusion: A Comparative Study. *J. Clin. Diagn. Res.* **2017**, *11*, ZC29–ZC33. [CrossRef]
36. Paknahad, M.; Shahidi, S.; Iranpour, S.; Mirhadi, S.; Paknahad, M. Cone-Beam Computed Tomographic Assessment of Mandibular Condylar Position in Patients with Temporomandibular Joint Dysfunction and in Healthy Subjects. *Int. J. Dent.* **2015**, *2015*, 301796. [CrossRef]
37. Guerrero Aguilar, A.; Flores Araque, M.E.; Flores Carrera, E.; Velásquez Ron, B. Condylar position and Temporomandibular Joint Space analysis with Cone-beam Tomography. *Odontol. Vital* **2021**, *2*, 6–16.
38. Lelis, E.R.; Guimarães Henriques, J.C.; Tavares, M.; de Mendonça, M.R.; Fernandes Neto, A.J.; Almeida, G.D.A. Cone-beam tomography assessment of the condylar position in asymptomatic and symptomatic young individuals. *J. Prosthet. Dent.* **2015**, *114*, 420–425. [CrossRef]
39. Choi, B.-J.; Kim, B.-S.; Lim, J.-M.; Jung, J.; Lee, J.-W.; Ohe, J.-Y. Positional change in mandibular condyle in facial asymmetric patients after orthognathic surgery: Cone-beam computed tomography study. *Maxillofac. Plast. Reconstr. Surg.* **2018**, *40*, 13. [CrossRef]
40. Chen, Y.; Wang, J.; Li, Y.; Li, L.; Luo, N.; Wu, Y.; Dai, H.; Zhou, J. Age-related variations in position and morphology of the temporomandibular joint in individuals with anterior openbite and crossbite: A multi-cross-sectional comparative study. *BMC Oral Health* **2022**, *22*, 200. [CrossRef]
41. Al-Hadad, S.A.; Alyafrusee, E.S.; Abdulqader, A.A.; Al-Gumaei, W.S.; Al-Mohana, R.A.A.M.; Ren, L. Comprehensive three-dimensional positional and morphological assessment of the temporomandibular joint in skeletal Class II patients with mandibular retrognathism in different vertical skeletal patterns. *BMC Oral Health* **2022**, *22*, 149. [CrossRef] [PubMed]
42. Ponces, M.J.; Tavares, J.P.; Lopes, J.D.; Ferreira, A.P. Comparison of condylar displacement between three biotypological facial groups by using mounted models and a mandibular position indicator. *Korean J. Orthod.* **2014**, *44*, 312–319. [CrossRef] [PubMed]
43. Almaqrami, B.S.; Alhammadi, M.S.; Tang, B.; Alyafrusee, E.S.; Hua, F.; He, H. Three-dimensional morphological and positional analysis of the temporomandibular joint in adults with posterior crossbite: A cross-sectional comparative study. *J. Oral Rehabil.* **2021**, *48*, 666–677. [CrossRef] [PubMed]
44. Westesson, P.L.; Bifano, J.A.; Tallents, R.H.; Hatala, M.P. Increased horizontal angle of the mandibular condyle in abnormal temporomandibular joints. A magnetic resonance imaging study. *Oral Surg. Oral Med. Oral Pathol.* **1991**, *72*, 359–363. [CrossRef]
45. Hegde, S.; Bn, P.; Shetty, S. Morphological and Radiological Variations of Mandibular Condyles in Health and Diseases: A Systematic Review. *Dentistry* **2013**, *3*, 1.
46. Karatas, O.H.; Toy, E. Three-dimensional imaging techniques: A literature review. *Eur. J. Dent.* **2014**, *8*, 132–140. [CrossRef]
47. Osiewicz, M.A.; Lobbezoo, F.; Loster, B.W.; Loster, J.E.; Manfredini, D. Frequency of temporomandibular disorders diagnoses based on RDC/TMD in a Polish patient population. *Cranio* **2018**, *36*, 304–310. [CrossRef]

Article

The Effect of TMJ Intervention on Instant Postural Changes and Dystonic Contractions in Patients Diagnosed with Dystonia: A Pilot Study

Ayse Selenge Akbulut

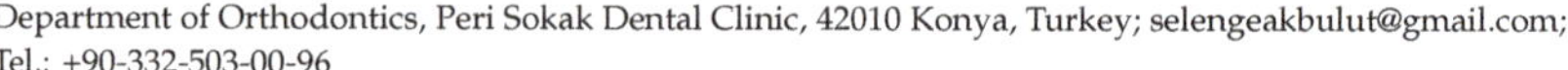

Department of Orthodontics, Peri Sokak Dental Clinic, 42010 Konya, Turkey; selengeakbulut@gmail.com; Tel.: +90-332-503-00-96

Abstract: (1) Background: The aim of the present study is to analyze the instant postural changes and changes in the dystonic contractions among patients with dystonia following the use of an intraoral device called a key integrative dynamic TMJ treatment appliance (KIDTA). (2) Methods: Twelve subjects, previously diagnosed with dystonia were enrolled. Their existing records were utilized to assess the changes in their posture and dystonic contractions. The posture analysis was conducted using a mobile application (APECS). The initial records (T0) and records acquired after the delivery of the KIDTA (T1) were utilized in the analysis. The Wilcoxon signed-rank test was performed to compare parameters between T0 and T1, with a significance level set at $p < 0.05$. (3) Results: Based on the Wilcoxon signed-rank test, statistically significant differences in T1 compared to T0 were observed in the severity of dystonic spasms, body alignment, head shift, head tilt, shoulder alignment, shoulder angle, axillae alignment, ribcage tilt, pelvic tilt, knee angle, and tibia angle ($p < 0.05$). (4) Conclusions: Within the limitations of the present pilot study, an intervention to the TMJ through a KIDTA appliance seems to mitigate the severity of dystonic contractions and improve the posture with respect to certain postural parameters.

Keywords: APECS; bruxism; cervical dystonia; dystonia; IDTT; orthodontic; posture; TMD; TMJ

Citation: Akbulut, A.S. The Effect of TMJ Intervention on Instant Postural Changes and Dystonic Contractions in Patients Diagnosed with Dystonia: A Pilot Study. *Diagnostics* **2023**, *13*, 3177. https://doi.org/10.3390/diagnostics13203177

Academic Editor: Luis Eduardo Almeida

Received: 16 September 2023
Revised: 4 October 2023
Accepted: 10 October 2023
Published: 11 October 2023

1. Introduction

The temporomandibular joint (TMJ) is a synovial joint consisting of the mandibular condyle and the mandibular fossa, which is located in the temporal bone. An articular disc, a tough fibrous connective tissue composed of compressed collagen fibers, is positioned between the mandibular fossa and mandibular condyle. This articular disc is indispensable for condylar regeneration and may serve as a crucial environmental factor for periosteal activation [1]. The proper functioning of the TMJ and its related components is crucial for directing the movement of the mandible and distributing the stresses that arise from routine activities like chewing, swallowing, talking, yawning, and other functions. On the other hand, parafunctional habits such as clenching, bruxism, pen chewing, nail biting, jaw thrusting, one-sided chewing, and repetitive or continuous external forces that create mandibular deviation in daily life might contribute to the development of temporomandibular disorder (TMD) and related degenerative changes [2–4]. Trauma, similar to parafunctional habits, is another factor that causes functional overloading in the TMJ region and contributes to the development of TMD. There is no consensus about the contribution of malocclusion to the development of TMD [5–8]. Although a previous systematic review referred to the absence of an association between TMD and dental occlusion [5], a high prevalence of TMD was reported in subjects with a dentofacial deformity in a previous meta-analysis [6]. The improper positioning of the mandibular condyle in the articular fossa can create stress in the TMJ complex and, as a result, affect the muscles, fascia, and neural network associated with the TMJ.

On the other hand, dystonia is a neurological condition characterized by involuntary, repetitive, and excessive muscular contractions. These contractions lead to abnormal movements, making it challenging for individuals to control their motions. The resulting movements and incorrect postures can be painful and affect the daily life of the individual. The prevalence of idiopathic or inherited dystonia was reported as 30.85 per 100,000 in a previous meta-analysis [9]. Dystonia was classified based on clinical characteristics and etiological factors by Grütz and Klein [10]. According to the clinical characteristics, dystonia was classified based on age, distribution area in the body, temporal pattern, and associated features. Based on the etiology, dystonia was classified under subcategories such as nervous system pathology, inherited or acquired, and others. Dystonia can impact various parts of the body, such as the neck, face, eyelids, jaw, vocal cords, hands, torso, and feet. Moreover, the repetitive muscle contractions in the affected body parts can lead to a fixed incorrect body posture. Correct posture aims to achieve maximum stability, conserve energy, and minimize stress on anatomical structures, but in patients with dystonia, incorrect posture makes it difficult to achieve these goals [11,12]. The body's position in space, which is automatically maintained through muscle contractions in response to gravity, can be associated with the TMJ [13–15]. Healthy and balanced TMJs are important not only for better chewing ability but also for better posture.

Integrative dynamic TMJ treatment (IDTT) is a treatment protocol developed by the author of the current study aimed at rehabilitating the TMJ by reducing the overloading on the TMJ, releasing probable stress in the relevant neural networks, and providing relaxation in the related muscles in close proximity. This protocol consists of two phases. In the first phase, the aim is to observe whether any overloading in the TMJ exists and whether a problem in the TMJ can be a risk factor for any other disease. These observations are carried out using a key appliance called the key integrative dynamic TMJ appliance (KIDTA) in the initial phase of the treatment protocol. If a positive correlation between the TMJ and symptoms of the disease is identified, the second phase becomes applicable. The second phase comprises multiple active treatment sessions in which integrative medicine techniques and physiotherapy techniques are employed, alongside the utilization of the integrative dynamic TMJ appliance (IDTA).

The aim of the current study is to analyze the acute postural changes and changes in the dystonic contractions in patients with dystonia after the instantaneous repositioning of the TMJ through an intraoral device called the KIDTA. Thus, the study aims to elucidate the effect of increasing the TMJ space on the body posture. The null hypothesis suggests that the immediate repositioning of the TMJ using the KIDTA may not result in alterations in posture and dystonic contractions when observed from both the frontal and lateral perspectives.

2. Materials and Methods

The protocol of this retrospective pilot study was approved by the ethical committee of the Department of Medicine at Necmettin Erbakan University (2023/4418). Initial records of subjects referred to Peri Sokak Dental Clinic (Konya, Turkey) between 2022 and 2023 were consecutively recruited.

Inclusion criteria:

- Subjects between the ages of 18 and 75;
- Subjects who have referred to the clinic with complaints of bruxism and/or myofascial pain;
- Subjects who have received a diagnosis of dystonia by a neurology specialist;
- Subjects who have started bruxism and TMJ treatment with the first phase of IDTT;
- Subjects with complete records related to postural assessment before and after intervention.

Exclusion criteria:

- Congenital anomaly;
- Incomplete photograph and video records;
- Subjects who received botox treatment within the last 4 months;

- Subjects who started or quit a new medication for the treatment of dystonia within 1 month.

According to the inclusion and exclusion criteria, 12 Caucasian subjects were included in this retrospective archive study.

KIDTA is a soft occlusal appliance made with Functional Impression Tissue Toner (FITT, Kerr, Italy) (Figure 1).

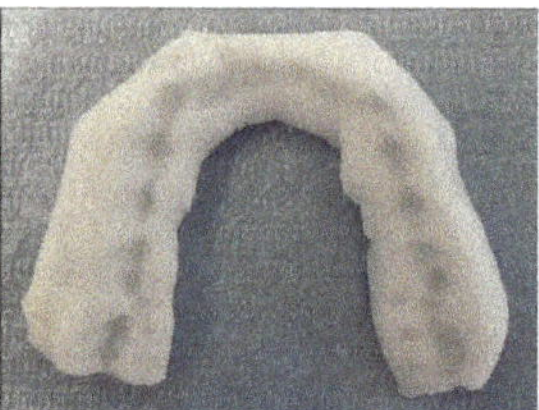

Figure 1. Key Integrative Dynamic Temporomandibular Joint Appliance (KIDTA).

Following the production of the dental models, the transition to the articulator is facilitated through bite registration. The bite registration wax is prepared based on an opening in the anterior region for 3 mm using Fleximeter Strips (Bausch flexi strips, Cologne, Germany). While biting the strips placed between the upper and lower incisors, a slight natural sliding movement in the anterior direction of up to 1 mm is allowed. KIDTA is an appliance that covers both the upper and lower occlusal surfaces. The appliance, which is fabricated using FITT, also covers approximately 1–3 mm of the teeth from the buccal surface. This soft material maintains its form for a duration of up to 10 days, necessitating the potential for periodic redelivery of the KIDTA appliance within the initial phase of treatment.

Postural analysis was performed on previously obtained photos and videos at two timepoints. T0 represented the time before any intervention with KIDTA, while T1 represented the time immediately after the delivery of KIDTA without any long-term usage.

2.1. Measurement Method

The photos and videos of the subjects that were acquired before and after intervention were collected. Postural analysis was performed on these photos using an application called APECS-AI Posture Evaluation and Correction System® (APECS mobile application) (New Body Technologies SAS, Grenoble, France). Postural evaluation was conducted in both frontal and lateral views. The landmarks used in postural analysis are provided in Table 1 for frontal view and Table 2 for lateral view.

Table 1. Descriptions and abbreviations of landmarks used in frontal view.

Landmark	Abbreviation	Description
Lobulus auriculae	LA	The lowest part of the ear.
Nose tip	NT	The tip of the nose.
Acromion	AC	The outer end of the scapula, extending over the shoulder joint and forming the highest point of the shoulder to which the clavicle is attached.
Jugular notch	JN	The largest visible dip in between the neck and the two clavicles in the superior margin of the sternum.
Anterior axillary fold	AAF	The ridge of the skin-covered inferior border of the pectoralis major muscle along the anterior sides of the chest where the underside of each arm meets the shoulder.

Table 1. *Cont.*

Landmark	Abbreviation	Description
Xiphoid process	XP	The pointed process of the cartilage at the lowest part of the sternum.
Lowest point of costal margin	LPCM	The most intended point on the contour of the torso inferiorly.
Most intended point of the trunk	MIPT	The most intended point on the contour of the torso laterally.
ASIS	ASIS	The prominence at the anterior projection of the iliac crest.
Anterior tibial tuberosity	ATT	The large oblong elevation on the proximal anterior aspect of the tibia.
Midpoint between malleoli	MBM	A midpoint of the line joining the most prominent bone on the outer side of the ankle and the most prominent bone on the inner side of the ankle.
Second metatarsophalangeal joint	2nd MTP	The joint between the metatarsal bone of the foot and the second proximal phalange.

Table 2. Descriptions and abbreviations of landmarks used in lateral view.

Landmark	Abbreviation	Description
Tragus	TG	A small pointed prominence of the external ear, situated in front of the ear canal.
C7 vertebra	C7	The most visible protrusion at the base of the neck.
Acromion	AC	The outer end of the scapula, extending over the shoulder joint and forming the highest point of the shoulder, to which the clavicle is attached.
ASIS	ASIS	The prominence at the anterior projection of the iliac crest.
PSIS	PSIS	The dimples between the buttocks and waist.
Greater trochanter	GT	A large protrusion located toward the top part of the shaft of the femur, facing outward.
Lateral joint	LJ	A slightly rounded enlargement on the surface of the knee.
Lateral malleolus	LM	The most prominent bone on the outer side of the ankle.
Head of the 5th metatarsal bone	5th MTH	A point at the prominence of the long bone (palpable along the distal outer edges of the feet) that connects the fifth proximal phalange to tarsal bone.

The images depicting the landmarks are provided in Figure 2 for the frontal view and Figure 3 for the lateral view.

In the frontal view, the vertical reference line was described as a line passing through the midpoint between the MBM points on the right and left sides, perpendicular to the ground surface. In the lateral view, the vertical reference line was determined as the line passing through the LM point and is perpendicular to the ground plane. The horizontal reference line for both planes was described as a line parallel to the ground surface.

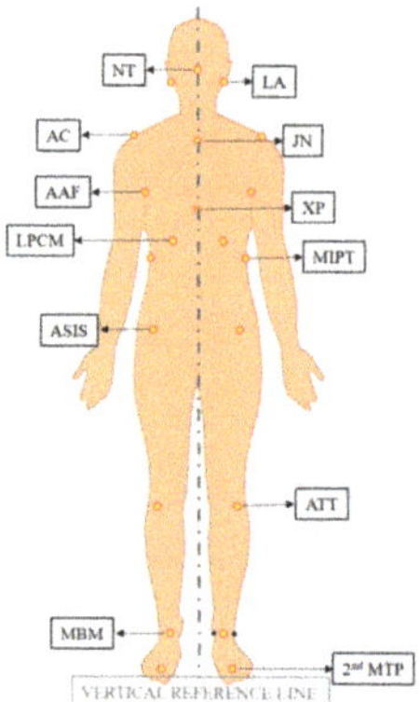

Figure 2. Landmarks used in frontal view.

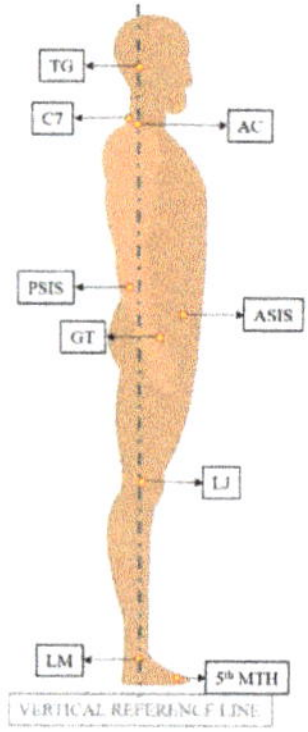

Figure 3. Landmarks used in lateral view.

The parameters measured in the frontal view and lateral view are detailed in Tables 3 and 4 for posture analysis, along with their corresponding abbreviations.

Table 3. Parameters measured in frontal view.

Parameter	Abbreviation	Description
Body Alignment Frontal	BAF	The angle formed between the vertical reference line and the line passing through the JN and the midpoint between the MBM points on the right and left sides.
Head Tilt	HT	The angle formed by the horizontal reference plane and the line joining the LA right and the LA left.
Shoulder Alignment	SAF	The angle formed by the horizontal reference plane and the line joining the AC right and the AC left.
Axillae Alignment	AA	The angle formed by the horizontal reference plane and the line joining the AAF right and the AAF left.
Ribcage Tilt	RT	The angle formed by the horizontal reference plane and the line joining the LPCM right and the LPCM left.
Most intended point of the trunk Tilt	MIPTT	The angle formed by the horizontal reference plane and the line joining the MIPT right and the MIIPT left.

Table 3. *Cont.*

Parameter	Abbreviation	Description
Pelvic Tilt Frontal	PTF	The angle formed by the horizontal reference plane and the line joining the ASIS right and the ASIS left.
Right Knee Angle	KAR	The acute angle formed between the line connecting the ASIS right and the ATT right, as well as the line connecting the ATT right and the MBM right.
Left Knee Angle	KAL	The acute angle formed between the line connecting the ASIS left and the ATT left, as well as the line connecting the ATT left and the MBM left.
Right Foot Rotation	FRR	The angle between the vertical reference line passing through the MBM right and the line connecting the MBM right and the 2nd MTP right.
Left Foot Rotation	FRL	The angle between the vertical reference line passing through the MBM left and the line connecting the MBM left and the 2nd MTP left.

Table 4. Parameters measured in lateral view.

Parameter	Abbreviation	Description
Body Alignment Lateral	BAL	The angle between the vertical reference line and the line connecting the LM and TG points.
Head Shift	HS	The angle between the vertical reference line and the line connecting the C7 and TG points.
Shoulder Angle	SAL	The angle between the horizontal reference line and the line connecting the C7 and AC points..
Pelvic Tilt Lateral	PTL	The acute angle between the horizontal reference line and the line connecting the ASIS and PSIS points.
Knee Angle	KA	The acute angle between the vertical reference line and the line connecting the GT and LJ points.
Tibia Angle	TA	The acute angle between the vertical reference line and the line connecting the LJ and LM points.
Foot Angle	FA	The angle between the horizontal reference line and the line connecting the LM and 5th MTH points.

The severity of dystonic contraction (SDC) was measured using video records. To calculate the SDC, the number of dystonic contractions per minute was counted. The duration of each individual dystonic contraction was measured in seconds. The measurement of the SDC was achieved by multiplying the total count of dystonic contractions by their respective durations.

Half of the total dataset was re-measured by another investigator to assess inter-rater reliability. To assess inter-rater reliability, an intra-class correlation coefficient was calculated. The intra-class correlation coefficient was above 0.90 which refers to a high reliability between two investigators (Table 5). In the final dataset, only the measurements from the first investigator were utilized.

Table 5. Inter-examiner reliability.

	Intra-Class Correlation	95% Confidence Interval		F Test with True Value 0	
		Lower Bound	Upper Bound	Value	df1
Average Measures	90.4%	0.860	0.935	10.434	107

2.2. Statistical Analysis

The statistical analysis of the data was performed using IBM SPSS Statistics Version 26.0 (Chicago, IL, USA). The statistical analysis was carried out through the Wilcoxon signed-rank test for comparing the measurements from before and after the intervention to the TMJ. The Wilcoxon signed-rank test was chosen due to the sample size of 12 participants in the present study, which falls below the conventional threshold of 30 for normality assumptions. According to the guidelines proposed by Sheskin (2003), non-parametric tests like the Wilcoxon signed-rank test are recommended when dealing with small sample sizes, as they can provide reliable inferences without necessitating stringent distributional requirements [16].

3. Results

The total sample consisted of eight females and four males, with an age range between 26 and 60 years. The mean values for the age, height, and weight were 41.5, 167.2, and 68.1, respectively. The demographic data of the subjects are provided in Table 6.

Table 6. Demographic Data.

	N	%	Age			Height		Weight	
			Range	Mean	S.D	Mean	S.D.	Mean	S.D.
Female	8	66.67%	26–60	42.8	12.8	164	4.5	66.0	13.2
Male	4	33.33%	26–49	39	10	173.5	2.6	72.3	6.1
Total	12	100%	26–60	41.5	11.6	167.2	6	68.1	11.4

S.D. Standard Deviation.

Among the 12 subjects, 10 were previously diagnosed with cervical dystonia, 1 subject was diagnosed with oromandibular dystonia, and 1 subject was diagnosed with writer's cramp.

The descriptive statistics for all parameters are provided in Table 7.

Table 7. Descriptive statistics for all parameters.

	N	Range	Minimum	Maximum	Mean		S.D.
					Statistic	Std. Error	Statistic
Severity of Dystonic Contraction-SDC-Pre	12	60.00	0.00	60.00	45.67	5.47	18.97
Severity of Dystonic Contraction-SDC-Post	12	26.00	0.00	26.00	12.17	2.53	8.78
Body Alignment Frontal-BAF-Pre	12	2.00	1.00	3.00	1.42	0.19	0.67
Body Alignment Frontal-BAF-Post	12	1.00	0.00	1.00	0.08	0.08	0.29
Head Tilt-HT-Pre	12	24.00	1.00	25.00	9.92	2.04	7.06
Head Tilt-HT-Post	12	4.00	0.00	4.00	1.25	0.41	1.42
Shoulder Alignment-SAF-Pre	12	9.00	0.00	9.00	2.83	0.73	2.52
Shoulder Alignment-SAF-Post	12	2.00	0.00	2.00	0.33	0.19	0.65
Axillae Alignment-AA-Pre	12	10.00	0.00	10.00	2.75	0.84	2.90
Axillae Alignment-AA-Post	12	2.00	0.00	2.00	0.58	0.19	0.67

Table 7. *Cont.*

	N	Range	Minimum	Maximum	Mean		S.D.
Ribcage Tilt-RT-Pre	12	6.00	0.00	6.00	1.50	0.49	1.68
Ribcage Tilt-RT-Post	12	3.00	0.00	3.00	0.83	0.32	1.12
Most intended point of the trunk Tilt-MIPTT-Pre	12	3.00	0.00	3.00	0.83	0.32	1.12
Most intended point of the trunk Tilt-MIPTT-Post	12	1.00	0.00	1.00	0.42	0.15	0.52
Pelvic Tilt Frontal-PTF-Pre	12	5.00	0.00	5.00	2.33	0.50	1.72
Pelvic Tilt Frontal-PTF-Post	12	2.00	0.00	2.00	0.83	0.24	0.84
Right Knee Angle-KAR-Pre	12	11.00	1.00	12.00	6.75	0.85	2.96
Right Knee Angle-KAR-Post	12	11.00	1.00	12.00	6.67	0.87	3.03
Left Knee Angle-KAL-Pre	12	11.00	3.00	14.00	6.75	1.03	3.57
Left Knee Angle-KAL-Post	12	10.00	3.00	13.00	6.50	0.10	3.45
Right Foot Rotation-FRR-Pre	12	36.00	−8.00	28.00	5.75	2.44	8.43
Right Foot Rotation-FRR-Post	12	21.00	0.00	21.00	5.92	1.60	5.53
Left Foot Rotation-FRL-Pre	12	25.00	−7.00	18.00	5.92	2.15	7.44
Left Foot Rotation-FRL-Post	12	21.00	0.00	21.00	6.25	1.89	6.54
Body Alignment Lateral-BAL-Pre	12	7.00	1.00	8.00	3.58	0.65	2.23
Body Alignment Lateral-BAL-Post	12	3.00	0.00	3.00	1.00	0.30	1.05
Head Shift-HS-Pre	12	80.00	29.00	109.00	49.75	6.18	21.41
Head Shift-HS-Post	12	26.00	21.00	47.00	36.17	2.06	7.13
Shoulder Angle-SAL-Pre	12	53.00	6.00	59.00	23.83	4.40	15.23
Shoulder Angle-SAL-Post	12	47.00	18.00	65.00	39.42	4.00	13.85
Pelvic Tilt Lateral-PTL-Pre	12	12.00	9.00	21.00	13.25	1.03	3.57
Pelvic Tilt Lateral-PTL-Post	12	11.00	6.00	17.00	11.75	0.91	3.14
Knee Angle-KA-Pre	12	6.00	0.00	6.00	3.42	0.57	1.98
Knee Angle-KA-Post	12	5.00	0.00	5.00	2.92	0.48	1.68
Tibia Angle-TA-Pre	12	10.00	1.00	11.00	5.75	0.83	2.86
Tibia Angle-TA-Post	12	9.00	0.00	9.00	4.75	0.77	2.67
Foot Angle-FA-Pre	12	22.00	16.00	38.00	28.25	1.92	6.65
Foot Angle-FA-Post	12	18.00	17.00	35.00	27.58	1.75	6.05

S.D.: Standard Deviation. Pre: measurement before intervention in T0. Post: measurement after intervention T1.

The results of the Wilcoxon signed-rank test are presented in Table 8 for the severity of the dystonic contractions, the postural measurements from the frontal view, and the postural measurements from the lateral view.

Table 8. The comparison of the parameters through the Wilcoxon signed-rank test.

		Total N	Test Statistic	Standard Error	Standardized Test Statistic	Asymptotic Sig. (2-Sided Test)
	Severity of Dystonic Contraction-SDC	12	0.000	11.247	−2.934	0.003 *
Frontal View	Body Alignment Frontal-BAF	12	0.000	10.909	−3.025	0.002 *
	Head Tilt-HT	12	0.000	11.242	−2.936	0.003 *
	Shoulder Alignment-SAF	12	0.000	9.753	−2.820	0.005 *
	Axillae Alignment-AA	12	2.000	9.753	−2.615	0.009 *
	Ribcage Tilt-RT	12	0.000	4.500	−2.333	0.020 *
	Most intended point of the trunk Tilt-MIPTT	12	2.000	3.623	−1.518	0.129
	Pelvic Tilt Frontal-PTF	12	0.000	9.657	−2.848	0.004 *
	Right Knee Angle-KAR	12	4.000	2.646	−0.378	0.705
	Left Knee Angle-KAL	12	0.000	1.732	−1.732	0.083
	Right Foot Rotation-FRR	12	4.000	2.739	−0.365	0.715
	Left Foot Rotation-FRL	12	13.500	7.045	−0.639	0.523
Lateral View	Body Alignment Lateral-BAL	12	0.000	11.164	−2.956	0.003 *
	Head Shift-HS	12	0.000	12.723	−3.065	0.002 *
	Shoulder Angle-SAL	12	66.000	11.247	2.934	0.003 *
	Pelvic Tilt Lateral-PTL	12	6.000	9.657	−2.226	0.026 *
	Knee Angle-KA	12	0.000	4.287	−2.449	0.014 *
	Tibia Angle-TA	12	0.000	8.147	−2.762	0.006 *
	Foot Angle-FA	12	8.000	8.016	−1.809	0.070

* Statistically significant difference ($p < 0.05$).

The parameter SDC exhibited a statistically significant difference between the measurements of T0 and T1 ($p < 0.05$).

From the frontal view parameters, BAF, HT, SAF, AA, RT, and PTF exhibited statistically significant differences between the measurements of T0 and T1 ($p < 0.05$). However, MIPTT, KAR, KAL, FRR, and FRL showed no statistically significant difference ($p > 0.05$).

Among the lateral view parameters, BAL, HS, SAL, PTL, KA, and TA exhibited statistically significant differences before and after the intervention ($p < 0.05$). However, only FA from the lateral view parameters did not show a statistically significant difference between the two timepoints ($p > 0.05$).

Representative videos of a subject are provided in Supplementary Materials. The alleviation of symptoms associated with dystonic contractions after TMJ intervention can be observed by comparing these before and after videos. Representative photos of two subjects in both the frontal and lateral views showing the alterations in body posture after the TMJ intervention, are presented in Figures 4 and 5.

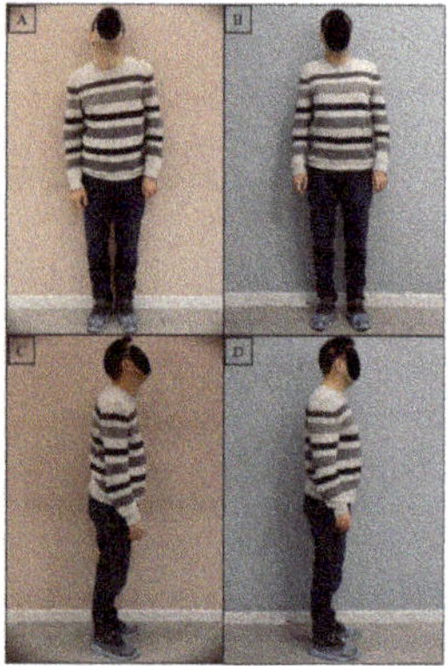

Figure 4. A representative photo of a subject presenting the alterations in body posture after TMJ intervention. (**A**) Posture in frontal view before TMJ intervention at T0, (**B**) posture in frontal view after TMJ intervention at T1, (**C**) posture in lateral view before TMJ intervention at T0, (**D**) posture in lateral view after TMJ intervention at T1.

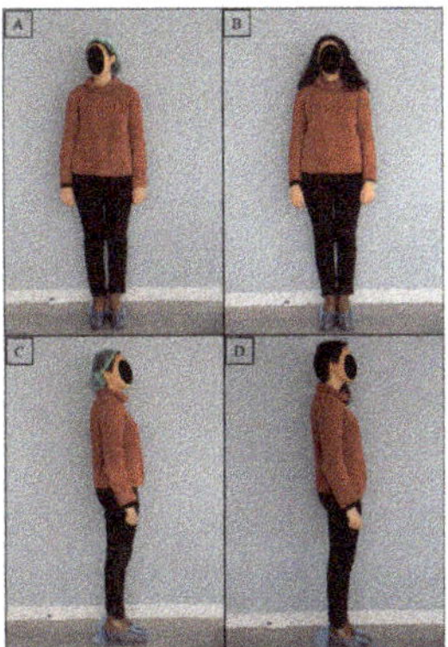

Figure 5. A representative photo of a subject presenting the alterations in body posture after TMJ intervention. (**A**) Posture in frontal view before TMJ intervention at T0, (**B**) posture in frontal view after TMJ intervention at T1, (**C**) posture in lateral view before TMJ intervention at T0, (**D**) posture in lateral view after TMJ intervention at T1.

4. Discussion

The results of the current study refer to the changes in certain parameters for both frontal and lateral assessments following intervention in the TMJ area through a KIDTA. These changes encompass not only the alterations in posture but also the alleviation of symptoms associated with dystonic contractions. In view of the findings from the present investigation, it is noteworthy that the null hypothesis was partially rejected.

The current study was designed as a retrospective and pilot study with a small sample size. Dystonia falls within the realm of neurology, and therefore, establishing a study group for this already relatively uncommon disease in the field of dentistry is anticipated to be quite challenging. Moreover, an overview of the information about a new treatment protocol that is under development has been provided in this pilot study. One of the appliances used in this treatment approach was presented with its impact on posture and dystonic contractions.

An incidence of bruxism between 28 and 34% in cervical dystonia and 17–74% in oromandibular dystonia cases was reported in previous studies [17–19]. Various kinds of questionnaires are being used for bruxism and TMD evaluation [20–22]. However, in the current study, only the subjects who were aware that they had bruxism and myofascial pain, and who were referred to the clinic with one of these chief complaints, were accepted as meeting the criteria for these issues.

Several mobile software applications that utilize artificial intelligence have emerged for the purpose of analyzing and evaluating posture. The PostureScreen® Mobile app is one of the most preferred applications used in previous studies [23,24]. However, extra effort is needed to understand which lines and planes the parameters are formed by. Therefore, the posture analysis was performed through the mobile application called the APECS mobile application in the present study. This application is a reliable tool that also utilizes artificial intelligence [25]. It presents how the relevant parameter is formed with specific points, lines, and planes, solely by analyzing the results. This approach enhances the efficiency of interpreting the analysis results. Moreover, the application provides detailed information about landmarks, contributing to the ease of applicability and simplicity. However, all measurements in this application are expressed as whole numbers. To provide more precise measurements, it is recommended that the program offers values with one or two decimal places after the comma.

The postural analysis and severity of dystonic contractions were evaluated using photos and videos obtained during routine IDTT intervention. Therefore, additional diagnostic tools such as EMG or radiologic images for postural assessment and evaluation of dystonic contractions were not available in this retrospective study. For the evaluation of the changes in dystonic contractions, a new parameter called the 'severity of dystonic contractions' was introduced in this pilot study. This parameter was calculated by multiplying the total number of dystonic contractions per minute by the mean value of duration in seconds for all the contractions that appeared in a minute. The maximum total score for the severity of the dystonic contractions was set at 60. Various characteristics of dystonic contractions were observed in the sample. Generally, each contraction lasted for a few seconds, and after a brief resting period, the contraction would start again, and so forth. In these cases, the severity of the dystonic contraction was measured as described, by counting the number of contractions and their duration. However, in some cases, there was only one contraction for the entire minute, which persisted continuously with a fixed head position until the patient intervened with their hands. In such instances, the number of contractions was recorded as one, while the duration was noted as 60 s. The characteristics of the dystonic contractions were also observed as tremors, which manifested as continuous shaking of the head but within a more limited range of motion compared to the previous dystonic contraction types. Due to the brief duration (possibly less than a second) of each tremor contraction and its continuous nature, the severity was determined as 60 for the subjects exhibiting tremors.

Although EMG evaluation would provide more accurate numerical values regarding the contractions, the parameter called the 'severity of dystonic contractions' could offer quantitative evaluation of video recordings in cases where EMG is unavailable. On the other hand, the severity of dystonia has previously been described using various scales [26]. Scales such as the Fahn–Marsden Rating Scale, the Bary Albright Dystonia Scale, and the Global Dystonia Severity Rating Scale are some of the scales that have been in use for evaluation of the overall severity of dystonia. These scales encompass the evaluations of affected body parts, triggering factors, related activities, and severity levels, among other factors. However, the severity of the dystonic contractions in the present study involved a simple calculation focusing solely on the contractions' duration and frequency, without accounting for affected body parts and other factors.

The posture evaluation was conducted using only two views instead of the usual four. The subjects' heads were positioned to either the left or right side. Many subjects exhibited a body twist and rotation towards one side. In individuals with conditions like kyphosis, lordosis, or sway back, one can anticipate similar postural evaluation outcomes for both left and right lateral assessments. Twisted and/or rotated postures rarely accompany these conditions. However, in subjects with dystonia, a rotated posture is expected to yield contrasting results for the right and left sides of the body during the posture analysis. To address this issue, an additional postural assessment from an axial view is considered necessary. Therefore, postural evaluation from the posterior and left lateral aspects has been reserved for future studies. Such studies would aim to investigate this matter further and utilize a three-dimensional posture analysis to provide a comprehensive explanation.

Various types of occlusal splints, produced through different techniques, have been previously discussed [27]. A previous study reported that both soft and hard occlusal splints have the capacity to alleviate TMJ symptoms [28]. Moreover, during a 4-month follow-up period, soft splints were observed to exhibit superiority over hard splints [28]. Although in some previous studies, the worsening of bruxism by the usage of soft appliances was reported [29,30], the findings of recent studies with objective methodologies offered opposite results [31–33]. An increase in the maximum bite force can be associated with bruxism [34]. In a recent study that used a digital gnathodynamometer [31], the maximum bite force was shown to increase when a hard splint was used and decrease when a soft splint was used. In another study where a portable electromyography (EMG) device was utilized, it was concluded that a soft occlusal splint was helpful for relieving pain owing to absorbing occlusal forces [32], whereas hard splints were found to increase muscle pain on palpation. Similarly, an increased bite force was reported for hard occlusal splints compared to soft occlusal splints in a recent study where the measurements were performed through

EMG [33]. The most significant characteristic of the KIDTA lies in the material employed during its fabrication. The soft material FITT is fabricated in an articulator rather than a vacuum forming machine. The KIDTA may absorb occlusal forces due to its soft nature, potentially alleviating stress in the TMJ. Moreover, using a soft material is considered as the most proper option when considering the dynamic nature of the overall treatment called IDTT. Rather than determining a proper position of the TMJ initially and keeping that position during treatment, the aim of KIDTA and IDTA is to find the balance of the TMJ during the treatment process of IDTT. Considering the movement of two different joints that are connected to each other, soft appliances could be helpful for the TMJ to find its balance in the treatment process owing to the dynamic changes in the distribution of forces.

The temporomandibular joint (TMJ) is an anatomically important area with close proximity to neural networks. It is richly innervated by sensory nerves, primarily branches of the trigeminal nerve. While the trigeminal nerve itself primarily carries sensory information, its interactions with other cranial nerves can influence motor functions and contribute to various motor disorders. The overload in the TMJ may significantly affect this neural system due to the crucial anatomical connections of the TMJ. By reducing and evenly distributing the load in the TMJ through a soft appliance, an enhancement of the neural system in this region could potentially occur. This mechanism has the potential to provide patients with the opportunity to facilitate their own healing by creating a healthier TMJ environment.

The severity of dystonic contractions decreased after wearing the KIDTA in the current study. Similar mechanisms were explored in other studies [35,36]. Symptom relief for dystonia was reported in a prior study involving the utilization of a hard occlusal stabilization appliance [35]. However, the methodological difference from the current study was the evaluation approach, as questionnaires were employed instead of the video records that were used here. Similarly, in a previous case report, an improvement in motor dysfunction related to Parkinson's disease (PD) was noted following the combined use of a bite splint alongside oral medication [36]. Another case report documented the enhancement of three subjects with cervical dystonia after utilizing an appliance called an orthotic [37].

In the frontal view, all parameters of the upper body showed a statistically significant difference except for the most intended point of the trunk tilt. However, only the pelvic tilt frontal showed a statistically significant difference among the parameters of the lower body from the frontal view.

A statistically significant decrease in the body alignment from the frontal view was observed in the current study. According to the measurement method in the current study, values closer to $0°$ refer to a well-aligned body from the frontal view [38]. A decrease in the mean value for the body alignment from $1.42°$ to $0.08°$ contributed to a better posture in the present study from the frontal view.

The frontal view parameters about tilts were measured through the lines formed by the connection of the bilateral identical points. The lobulus auriculae was the reference point for assessing the head tilt. A similar horizontal level of the lobulus auriculae points indicates a normal alignment [38]. The initial mean value of the head tilt was $9.92°$, indicating a severe head tilt in patients with dystonia. However, this mean value decreased to $1.25°$ after the intervention, approaching the normal range of $0–0.1°$ [38]. Our results for the head tilt were also compatible with the mean value of $2°$ reported in a previous study conducted on young, healthy individuals [39].

The shoulder alignment in the frontal view was measured based on the acromion point in the current study. The normal value for the shoulder alignment was reported as $181°$, which translates to $1°$ based on the current measurement method in a previous study [38]. The only difference was the reference point for the measurement, which was the coracoid process in the previous study. When assessing the tilt in the frontal view, symmetry is of the utmost importance. Therefore, even when different reference points were used in the measurements, similar norms could apply to the measurements performed based on close proximity. Considering a normal value of up to $1°$, the mean value of $2.83°$ also indicates a tilt in the shoulder in subjects with dystonia. However, the decrease of this value to 0.33° after the intervention reveals a correction in the shoulders as well. A shoulder alignment of 1.3° in a previous study was also in line with the current results for T1 [39].

A significant decrease in the axillae alignment, ribcage tilt, and pelvic tilt after the use of the KIDTA indicates a correction in these postural parameters. The mean values for these parameters (AA: 0.58°, RT: 0.83°, PT: 0.83°) in T1 were also consistent with the mean values of the previous study (AA: 1.3°, RT: 1.9°, PT: 2.3°) that utilized the same measurement method [39].

The trunk inclination was reported as 1.6° in a previous study in which young healthy adults were examined [39]. This result was consistent with the findings for the most intended point of the trunk tilt in both T0 and T1 parameters (MIPTT-Pre: 0.83°, MIPTT-Post: 0.42°). This similarity could explain the non-significant difference between them. Similarly, the frontal parameters concerning the knee angle did not show a statistically significant difference between T0 and T1, with mean values ranging between 6.75° and 6.50°. These values were also similar to those in a previous study that used the same landmarks [39].

All parameters showed a statistically significant difference in the lateral view parameters except for the foot angle. When compared to T0, the body alignment lateral, head shift, pelvic tilt, knee angle, and tibia angle were decreased in T1 in the lateral view. However, the shoulder angle increased in T1 compared to T0.

An increased value of the body alignment lateral in subjects with dystonia was observed in the present study (mean: 3.58°). However, intervention to the TMJ enabled the subjects to maintain a better posture from the lateral view with a better body alignment (mean: 1.00°).

The head shift was measured as the angle between the vertical reference line and the line connecting the C7 and TG points in the present study. A similar measurement was performed in previous studies as the angle between the horizontal line and the line connecting the tragus with C7 [38]. This measurement was named as the craniovertebral angle. An increase in the craniovertebral angle means a less forward head posture [40]. Therefore, the decreased values of the head shift in T1 (mean: 36.17°) compared to T0 (mean: 49.75°) indicate an enhancement in forward head posture in the present study. However, the head shift after the intervention was less than 31.4, which was reported in a previous study [38]. This can be explained by the study sample of the previous study, which consisted of healthy subjects. In another previous study, the mean craniovertebral angles ranged between 32.67° and 46.83°. These values referred to a more forward head posture compared to the T1 values. However, they were compatible with the results in T0. The mean age could explain this, with a mean age of 78.42 in the previous study and a mean age of 41.5 in the current study [41].

The measurement method of the shoulder angle was similar to the studies in the literature [38,40]. Lower values of the shoulder angle indicate a more kyphotic posture, forward head posture, and rounded shoulders [40]. The shoulder angle was increased in T1 (mean: 39.42°) compared to T0 (mean: 23.83°). This indicated an improvement in the shoulder angle. However, the values even after the intervention could not reach the shoulder angles of the other studies. Mean values of 51.4° and 53.7° were reported in previous studies where asymptomatic subjects were examined [38,42]. However, the results were higher than a previous study where a mean angle of 19.6° was reported [39]. According to the results of that previous study, the shoulder alignment was the only parameter in sagittal view that failed the reproducibility analysis.

In a previous study, it was suggested that an anterior pelvic tilt up to a certain degree could be a typical finding in asymptomatic subjects [43]. They reported mean degrees of the anterior pelvic tilt as 6.74° and 6.23° for the left and right sides, respectively, in males. For females, the values were reported as 6.93° for the left side and 6.63° for the right side [43]. In another study conducted on healthy adults and athletes, the mean values of the anterior pelvic tilt were reported as 9.6° and 11.7° for males and females, respectively [44].

Furthermore, in a study based on radiographic images, a mean anterior pelvic tilt value of 13° was reported [45]. In another study conducted on healthy college students, the mean anterior pelvic tilt values were reported as 8.6° (right) and 8.7° (left) for males, and 12.2° (right) and 11.8° (left) for females [46]. In a previous study, a postural assessment was conducted through a mobile application from the sagittal view [39]. The mean value of the pelvic tilt was reported as 16.9° in young healthy adults. A statistically significant decrease in the mean values of the pelvic tilt from the lateral view was observed in the current study after the intervention. The mean values for T0 (13.25°) and T1 (11.75°) could be evaluated as consistent with the previous studies [44–46]. However, the lack of consensus about the norms of pelvic tilt should be considered when evaluating the pelvic tilt.

A significant difference in the knee angle and tibia angle was also observed in the current study. Although the intervention did not affect the lower limbs in the frontal view, it caused a change in the legs when evaluated from the lateral view. This effect could be related to changes in the pelvic tilt as well as the enhancement of the overall posture in the lateral view. The relationship between the anterior pelvic tilt and lower extremity kinematics was presented in a previous study [47]. Although static posture was investigated in the current study, the association between the legs and the pelvic tilt could also have an impact on static posture as well.

The total sample size was the main limitation of the current study, which was conducted on 12 subjects. Due to the relative rarity of dystonia as a neurological disorder, obtaining a large sample size in the dental area is challenging. Therefore, the study was designed as a pilot study. Another limitation was related to the selection criteria. The subjects were not using the same medications. This could affect the results. It is believed that changing or discontinuing the medications that patients routinely use may have some effects on the posture and dystonic contractions. To avoid influencing the parameters due to any changes in medication, the medications were not altered, and the comparisons were performed between before and after the intervention within groups. For further studies, a sample that includes individuals undergoing the same medical interventions could be considered to eliminate the effects of factors other than the TMJ intervention.

The current pilot study presented the immediate changes that occurred in the static posture and dystonic contractions after intervention to the TMJ. An overview of a new therapeutic approach under development was also provided in the current study. The findings emphasize the importance of the TMJ not only for healthy individuals but also for patients with dystonia. By presenting the relationship between the TMJ and posture in the current study, the importance of the TMJ has once again been highlighted. Clinical suggestions and implementations of the current study could include taking preventative measures to maintain TMJ health, conducting detailed routine examinations of the TMJ, treating temporomandibular disorders with appropriate timing and methods, recommending the inclusion of routine TMJ evaluations in the assessment process for individuals with dystonia, and being open to multidisciplinary collaborations that consider a holistic approach to the human body.

5. Conclusions

Within the limitations of the present pilot study, an intervention in the TMJ position through the KIDTA can lead to immediate changes in the posture of subjects with dystonia, including the body alignment, head shift, head tilt, shoulder alignment, shoulder angle, axillae alignment, ribcage tilt, pelvic tilt, knee angle, and tibia angle. Additionally, this intervention could mitigate the severity of dystonic spasms in subjects with dystonia.

Supplementary Materials: The following supporting information can be viewed at: https://www.youtube.com/watch?v=eXL6R1EQBos (accessed on 9 September 2023), https://www.youtube.com/watch?v=qrd1GL4CHsM (accessed on 9 September 2023).

Funding: This research received no external funding.

Institutional Review Board Statement: The study was conducted in accordance with the Declaration of Helsinki, and was approved by the Ethics Committee of Necmettin Erbakan University Faculty of Medicine (protocol code: 2023/4325, date of approval: 5 May 2023).

Informed Consent Statement: Informed consent was obtained from all subjects involved in the current study.

Data Availability Statement: Research data are available on request due to privacy or ethical restrictions.

Acknowledgments: The author would like to express her gratitude to Gokcen Sahin and the neurologist Aysun Hatice Akca Karpuzoglu for their contributions to the current study. The author extends her sincere appreciation to Serife Bilgic and Turkay Kolus for their significant contributions to this study, particularly in the domain of statistical analysis.

Conflicts of Interest: The author declares no conflict of interest.

References

1. Hayashi, H.; Fujita, T.; Shirakura, M.; Tsuka, Y.; Fujii, E.; Terao, A.; Tanimoto, K. Role of articular disc in condylar regeneration of the mandible. *Exp. Anim.* **2014**, *63*, 395–401. [CrossRef] [PubMed]
2. Yadav, U.; Ahmed, J.; Ongole, R.; Shenoy, N.; Sujir, N.; Natarajan, S. Influence of psychosocial factors and parafunctional habits in temporomandibular disorders: A cross-sectional study. *Perm. J.* **2020**, *24*, 144. [CrossRef] [PubMed]
3. Sójka, A.; Stelcer, B.; Roy, M.; Mojs, E.; Pryliński, M. Is there a relationship between psychological factors and TMD? *Brain Behav.* **2019**, *9*, e01360. [CrossRef] [PubMed]
4. Santana-Mora, U.; López-Cedrún, J.; Suárez-Quintanilla, J.; Varela-Centelles, P.; Mora, M.J.; Da Silva, J.L.; Figueiredo-Costa, F.; Santana-Penín, U. Asymmetry of dental or joint anatomy or impaired chewing function contribute to chronic temporomandibular joint disorders. *Ann. Anat.* **2021**, *238*, 151793. [CrossRef]
5. Manfredini, D.; Lombardo, L.; Siciliani, G. Temporomandibular disorders and dental occlusion. A systematic review of association studies: End of an era? *J. Oral Rehabil.* **2017**, *44*, 908–923. [CrossRef]
6. Al-Moraissi, E.A.; Perez, D.; Ellis III, E. Do patients with malocclusion have a higher prevalence of temporomandibular disorders than controls both before and after orthognathic surgery? A systematic review and meta-analysis. *J. Craniomaxillofac. Surg.* **2017**, *45*, 1716–1723. [CrossRef]
7. Macrì, M.; Murmura, G.; Scarano, A.; Festa, F. Prevalence of temporomandibular disorders and its association with malocclusion in children: A transversal study. *Front. Public Health* **2022**, *10*, 860833. [CrossRef] [PubMed]
8. Manfredini, D.; Perinetti, G.; Guarda-Nardini, L. Dental malocclusion is not related to temporomandibular joint clicking: A logistic regression analysis in a patient population. *Angle Orthod.* **2014**, *84*, 310–315. [CrossRef]
9. Medina, A.; Nilles, C.; Martino, D.; Pelletier, C.; Pringsheim, T. The Prevalence of Idiopathic or Inherited Isolated Dystonia: A Systematic Review and Meta-Analysis. *Mov. Disord. Clin. Pract.* **2022**, *9*, 860–868. [CrossRef]
10. Grütz, K.; Klein, C. Dystonia updates: Definition, nomenclature, clinical classification, and etiology. *J. Neural Transm.* **2021**, *128*, 395–404. [CrossRef]
11. Albanese, A.; Di Giovanni, M.; Lalli, S. Dystonia: Diagnosis and management. *Eur. J. Neurol.* **2019**, *26*, 5–17. [CrossRef]
12. Pastorelli, F.; Pasquetti, P. Biomechanical analysis and rehabilitation in athletes. *Clin. Cases Miner Bone Metab.* **2013**, *10*, 96.
13. Salkar, R.G.; Radke, U.M.; Deshmukh, S.P.; Radke, P.M. Relationship between temporomandibular joint disorders and body posture. *Int. J. Dent. Health Sci.* **2015**, *2*, 1523–1530.
14. Fiorillo, L. Spine and TMJ: A Pathophysiology report. *J. Funct. Morphol. Kinesiol.* **2020**, *5*, 24. [CrossRef] [PubMed]
15. Fiorillo, L.; Musumeci, G. TMJ dysfunction and systemic correlation. *J. Funct. Morphol. Kinesiol.* **2020**, *5*, 20. [CrossRef] [PubMed]
16. Sheskin, D.J. *Handbook of Parametric and Nonparametric Statistical Procedures*, 3rd ed.; Chapman and Hall/CRC Press: London, UK, 2003; pp. 1–1193.
17. Paus, S.; Gross, J.; Moll-Müller, M.; Hentschel, F.; Spottke, A.; Wabbels, B.; Klockgether, T.; Abele, M. Impaired sleep quality and restless legs syndrome in idiopathic focal dystonia: A controlled study. *J. Neurol.* **2011**, *258*, 1835–1840. [CrossRef]
18. Watts, M.W.; Tan, E.K.; Jankovic, J. Bruxism and cranial-cervical dystonia: Is there a relationship? *Cranio* **1999**, *17*, 196–201. [CrossRef]
19. Singer, C.; Papapetropoulos, S. A comparison of jaw-closing and jaw-opening idiopathic oromandibular dystonia. *Park. Relat. Disord.* **2006**, *12*, 115–118. [CrossRef]
20. Machado, N.A.G.; Costa, Y.M.; Quevedo, H.M.; Stuginski-Barbosa, J.; Valle, C.M.; Bonjardim, L.R.; Garib, D.G.; Conti, P.C.R. The association of self-reported awake bruxism with anxiety, depression, pain threshold at pressure, pain vigilance, and quality of life in patients undergoing orthodontic treatment. *J. Appl. Oral Sci.* **2020**, *28*, e20190407. [CrossRef]
21. Schiffman, E.; Ohrbach, R. Executive summary of the Diagnostic Criteria for Temporomandibular Disorders for clinical and research applications. *J. Am. Dent. Assoc.* **2016**, *147*, 438–445. [CrossRef]
22. Lavigne, G.J.; Montplaisir, J.V. Bruxism: Epidemiology, diagnosis, pathophysiology, and pharmacology. In *Orofacial Pain and Temporomandibular Disorders: Advances in Pain Research and Therapy*, 1st ed.; Fricton, J.R., Dubner, R., Eds.; Raven Press: New York, NY, USA, 1995; Volume 21, pp. 387–404.

23. Iacob, S.M.; Chisnoiu, A.M.; Lascu, L.M.; Berar, A.M.; Studnicska, D.; Fluerasu, M.I. Is PostureScreen® Mobile app an accurate tool for dentists to evaluate the correlation between malocclusion and posture? *Cranio* **2020**, *38*, 233–239. [CrossRef] [PubMed]

24. Santos, J.G.L.; Montezuma, T.; Perez, C.S.; Sverzut, C.E.; Trivellato, A.E.; de Oliveira Guirro, E.C. Body postural realignment in the first 2 months after orthognathic surgery. *Am. J. Orthod. Dentofac. Orthop.* **2021**, *159*, e281–e290. [CrossRef] [PubMed]

25. Welling, A.; Gurudut, P.; Shirodkar, G.; Shetye, N.; Khan, S. Validation of non-radiographic APECS software in comparison with standard radiographic measurement of full-length lower limb hip-knee-ankle angle in elderly obese women. *Physiother. Q.* **2023**, *31*, 90–94. [CrossRef]

26. Comella, C.L.; Leurgans, S.; Wuu, J.; Stebbins, G.T.; Chmura, T.; Dystonia Study Group. Rating scales for dystonia: A multicenter assessment. *Mov. Disord.* **2003**, *18*, 303–312. [CrossRef]

27. Okeson, J.P. *Management of Temporomandibular Disorders and Occlusion*, 6th ed.; Elsevier: St. Louis, MO, USA, 1989.

28. Seifeldin, S.A.; Elhayes, K.A. Soft versus hard occlusal splint therapy in the management of temporomandibular disorders (TMDs). *Saudi Dent. J.* **2015**, *27*, 208–214. [CrossRef]

29. Okeson, J.P. The effects of hard and soft occlusal splints on nocturnal bruxism. *J. Am. Dent. Assoc.* **1987**, *114*, 788–791. [CrossRef]

30. Harkins, S.; Marteney, J.L.; Cueva, O.; Cueva, L. Application of soft occlusal splints in patients suffering from clicking temporomandibular joints. *Cranio* **1988**, *6*, 71–76. [CrossRef]

31. Benli, M.; Özcan, M. Short-term effect of material type and thickness of occlusal splints on maximum bite force and sleep quality in patients with sleep bruxism: A randomized controlled clinical trial. *Clin. Oral Investig.* **2023**, *27*, 4313–4322. [CrossRef]

32. Kolcakoglu, K.; Dogan, S.; Tulga Oz, F.; Aydınbelge, M. A comparison of hard and soft occlusal splints for the treatment of nocturnal bruxism in children using the BiteStrip®. *J. Clin. Pediatr. Dent.* **2022**, *46*, 219–224. [CrossRef]

33. Lei, Q.; Lin, D.; Liu, Y.; Lin, K.; Huang, W.; Wu, D. Neuromuscular and occlusion analysis to evaluate the efficacy of three splints on patients with bruxism. *BMC Oral Health* **2023**, *23*, 325. [CrossRef]

34. Todić, J.T.; Mitić, A.; Lazić, D.; Radosavljević, R.; Staletović, M. Effects of bruxism on the maximum bite force. *Vojn. Pregl.* **2017**, *74*, 138–144. [CrossRef]

35. Navrotchi, C.; Badea, M.E. The influence of occlusal stabilization appliances on cervical dystonia symptoms. *Clujul Med.* **2017**, *90*, 438–444. [CrossRef] [PubMed]

36. Nomoto, S.; Nakamura, M.; Sato, T.; Hisanaga, R. Occlusal treatment with bite splint improves dyskinesia in Parkinson's disease patient: A case report. *Bull. Tokyo Dent. Coll.* **2013**, *54*, 157–161. [CrossRef] [PubMed]

37. Sims, A.B.; Stack, B.C.; Demerjian, G.G. Spasmodic torticollis: The dental connection. *Cranio* **2012**, *30*, 188–193. [CrossRef]

38. Raine, S.; Twomey, L.T. Head and Shoulder Posture Variations in 160 Asymptomatic Women and Men. *Arch. Phys. Med. Rehabil.* **1997**, *78*, 1215–1223. [CrossRef]

39. Trovato, B.; Roggio, F.; Sortino, M.; Zanghì, M.; Petrigna, L.; Giuffrida, R.; Musumeci, G. Postural Evaluation in Young Healthy Adults through a Digital and Reproducible Method. *J. Funct. Morphol. Kinesiol.* **2022**, *7*, 98. [CrossRef]

40. Alkhateeb, A.M.; Daher, N.S.; Forrester, B.J.; Martin, B.D.; Jaber, H.M. Effects of adjustments to wheelchair seat to back support angle on head, neck, and shoulder postures in subjects with cerebral palsy. *Assist. Technol.* **2021**, *33*, 326–332. [CrossRef]

41. Nemmers, T.M.; Miller, J.W.; Hartman, M.D. Variability of the forward head posture in healthy community-dwelling older women. *J. Geriatr. Phys. Ther.* **2009**, *32*, 10–14. [CrossRef]

42. Ruivo, R.M.; Pezarat-Correia, P.; Carita, A.I. Cervical and shoulder postural assessment of adolescents between 15 and 17 years old and association with upper quadrant pain. *Braz. J. Phys. Ther.* **2014**, *18*, 364–371. [CrossRef]

43. Herrington, L. Assessment of the degree of pelvic tilt within a normal asymptomatic population. *Man. Ther.* **2011**, *16*, 646–648. [CrossRef]

44. Medina McKeon, J.M.; Hertel, J. Sex differences and representative values for 6 lower extremity alignment measures. *J. Athl. Train.* **2009**, *44*, 249–255. [CrossRef] [PubMed]

45. Vialle, R.; Levassor, N.; Rillardon, L.; Templier, A.; Skalli, W.; Guigui, P. Radiographic analysis of the sagittal alignment and balance of the spine in asymptomatic subjects. *J. Bone Jt. Surg. Am.* **2005**, *87*, 260–267. [CrossRef]

46. Nguyen, A.D.; Shultz, S.J. Sex differences in clinical measures of lower extremity alignment. *J. Orthop. Sports Phys. Ther.* **2007**, *37*, 389–398. [CrossRef] [PubMed]

47. Alizadeh, S.; Mattes, K. How anterior pelvic tilt affects the lower extremity kinematics during the late swing phase in soccer players while running: A time series analysis. *Hum. Mov. Sci.* **2019**, *66*, 459–466. [CrossRef] [PubMed]

 *diagnostics*

Article

Evaluation of Temporomandibular Joint in Patients with Parkinson's Disease: A Comparative Study

Ayse Selenge Akbulut [1],* and Aysun Hatice Akca Karpuzoglu [2]

[1] Department of Orthodontics, Peri Sokak Dental Clinic, 42010 Konya, Türkiye
[2] Department of Neurology, Medova Hospital, 42070 Konya, Türkiye; aysun.karpuzoglu@medova.com.tr
* Correspondence: selengeakbulut@gmail.com; Tel.: +90-332-503-00-96

Abstract: (1) The aim of this study was to perform an evaluation of the temporomandibular joint (TMJ) in patients with Parkinson's disease (PD) and present the morphological differences of the TMJ between healthy subjects and patients with PD. (2) A total of 102 Caucasian subjects were divided equally into two groups. The study group consisted of patients with PD, while the control group comprised healthy subjects. Ten parameters, including anterior joint space (AJS), superior joint space (SJS), posterior joint space (PJS), condyle head length (CHL), condylar neck width (CNW), minor axis of the condyle (MAC), long axis of the condyle (LAC), condylar axis inclination (CI), medial joint space (MJS), and lateral joint space (LJS), were measured using magnetic resonance images. The data were statistically analyzed using paired samples t-test and Student's t-test, with a significance level set at $p < 0.05$. (3) In the PD group, all TMJ parameters showed a statistically significant difference between both sides of the face ($p < 0.05$). However, in the control group, AJS, SJS, PJS, CHL, CNW, MAC, CI, MJS, and LJS did not show a statistically significant difference between both sides of the face ($p > 0.05$), except for LAC ($p < 0.05$). The asymmetry index values of AJS, SJS, PJS, CHL, CNW, MAC, CI, MJS, and LJS demonstrated a statistically significant difference between the study and control groups ($p < 0.05$), except for LAC ($p > 0.05$). (4) Within the limitations of this retrospective study, the findings suggest that TMJ morphology and asymmetry could be associated with PD.

Keywords: MR; orthodontics; Parkinson's disease; TMD; TMJ

Citation: Akbulut, A.S.; Akca Karpuzoglu, A.H. Evaluation of Temporomandibular Joint in Patients with Parkinson's Disease: A Comparative Study. *Diagnostics* **2023**, *13*, 2482. https://doi.org/10.3390/diagnostics13152482

Academic Editor: Luis Eduardo Almeida

Received: 31 May 2023
Revised: 11 July 2023
Accepted: 19 July 2023
Published: 26 July 2023

1. Introduction

The temporomandibular joint (TMJ) is a specific, complex bilateral joint that connects the mandible to the temporal bone. Clinicians deal with various symptoms of temporomandibular disorders (TMDs), such as myofascial pain, TMJ sounds, restriction of mandibular movements, headache, or ear problems, in daily practice [1,2]. The prevalence of TMDs in adults and the elderly has been reported as 31.1% in a recent meta-analysis [3]. Considering this important prevalence rate of TMDs, the etiological factors underlying TMDs have also been gaining importance. Occlusal factors, psychological factors, hormonal factors, microtrauma, parafunctional habits, joint hyperlaxity and joint hypermobility, and hereditary factors are some of the etiological factors that take part in the occurrence of TMD [4].

Parkinson's disease (PD) is a neurodegenerative disorder that affects approximately 6.1 million people worldwide [5], and its incidence is growing due to the increasing aging of the population. The exact cause of PD is not completely known, but it is recognized that a decline in dopamine levels occurs due to the degeneration of dopaminergic neurons in the substantia nigra [6]. While a cure for PD is not presently accessible, it is feasible to alleviate the symptoms by employing dopaminergic replacement therapy. Levodopa, which acts as a precursor to dopamine, is frequently employed in the medical management of PD symptoms [7]. Some specific gene mutations can be attributed to PD in approximately 5% to 10% of cases. However, the majority of individuals with PD do not possess these mutations [8,9]. This may indicate that PD may also be involved with undiscovered genetic mutations in addition to known genetic mutations, as well as other environmental and lifestyle-related factors, such as head trauma, exposure to chemical toxins, and smoking [9]. Environmental and genetic factors influence a shared network of pathways that involve oxidative stress, mitochondrial dysfunction, protein aggregation, neuroinflammation, and impaired autophagy [10,11].

Some of the motor and nonmotor symptoms of PD are bradykinesia, muscle stiffness, tremor, body imbalance, impaired posture and coordination, freezing of gate, constipation, fatigue, anxiety, cognitive impairment, depression, and sleeping disorders [12–17]. The presence of symptoms such as rigidity, muscle stiffness, and tremor in the masticatory muscles in PD can result in clenching, bruxism, and TMDs [18,19]. Additionally, given that PD predominantly affects the motor system, it is likely that the resulting motor symptoms directly contribute to the development of TMDs.

On the other hand, the complex neural network in the TMJ area prompts researchers to question its potential relationship with certain neurodegenerative diseases. Sensory signals from the TMJ are transmitted through the trigeminal nerve to the brainstem, where they are processed and relayed to higher brain regions responsible for pain perception and motor control. Motor neurons in the brainstem play a crucial role in the control of movement throughout the body, and their dysfunction can have significant implications in various neurological diseases [20,21]. From this perspective, instead of considering TMD solely as a symptom of PD, it is essential to investigate the TMJ to determine whether it can serve as an underlying biomechanical risk factor for PD.

Previous studies have focused on the relationship between TMJ and PD [22–26]. However, considering the important anatomical connections of TMJ, more studies that investigate the relationship between TMJ and PD on a morphological basis are needed for better understanding.

This study aimed to perform an evaluation of the TMJ in patients with PD and present the morphological differences of TMJ between healthy subjects and patients with PD. The null hypotheses were as follows:

- Patients with PD show symmetry in their bilateral temporomandibular joints;
- Healthy individuals show asymmetry in their bilateral temporomandibular joints;
- There is no difference in temporomandibular joint symmetry between healthy individuals and individuals with PD.

2. Materials and Methods

The protocol of this retrospective split-mouth study was approved by the ethical committee of Necmettin Erbakan University (2023/4325). Initial records of subjects referred to Medova Hospital, Department of Neurology (Konya, Turkey) between 2017 and 2023 were consecutively recruited.

Patients diagnosed with PD, patients between 40 and 95 years of age, and patients who had magnetic resonance (MR) images of the brain that also included the TMJ area clearly were selected for the study group. The health management information system enables users to filter patients based on different categories. In this case, the filter settings were adjusted to select only patients diagnosed with PD between the years 2017 and 2023. Additional filtering functions were then employed to meet the criteria of having brain MR images and specific gender requirements.

Healthy subjects between 40 and 95 years of age and subjects who had MR images of the brain that included the TMJ area clearly were selected for the control group. For selection of control group, the filter settings of health management information system were adjusted to select only check-up subjects with no specific disease between the years 2017 and 2023. From this group, subjects who had undergone brain MR scans were selected using the filter function. The subjects were then categorized based on their gender.

Subjects with craniofacial anomalies, a menton deviation greater than 2 mm, a history of traumatic injury, or a history of surgery in the craniofacial region were excluded from both the study and control groups. Menton deviation was measured in the coronal view. The perpendicular distance from the menton point to the midsagittal plane (MSP) was measured. If the distance was more than 2 mm, the subject was excluded from the study. History of traumatic injury and history of surgery in the craniofacial region were assessed through a nationwide patient record system, and the health management information system was provided by the hospital.

According to the power analysis conducted using G* Power software (version 3.1; Heinrich Heine University, Düsseldorf, Germany), it was determined that a minimum of 102 participants would need to be included in the study, considering a power of 0.80 and an effect size of 0.5. For the study group, 51 Caucasian patients were randomly selected, and for the control group, another 51 Caucasian subjects were also selected randomly using random allocation software.

The study was divided into three parts. In the first part, a morphological comparison of the TMJs between both sides was performed in the study group. In the second part, a morphological comparison of the TMJs between both sides was performed in the control group. The third part involved a morphological and asymmetrical comparison of the TMJs between the study and control groups.

2.1. Measurement Method

In the current study, MR images of the brain were utilized. The MR images were captured using the same digital machine (Siemens Magnetom Aera 1.5 Tesla, Siemens Healthcare, Erlangen, Germany) and operated by the same individual. T2-weighted images were acquired using a standardized method during the MR image acquisition process. Measurements were conducted using the same software (Ea Medware Pacs Version 3.0.0.0., EA software, Turkey) by a single investigator. To assess inter-examiner reliability, the measurements for five randomly selected subjects from study and control groups were repeated by another examiner.

The TMJ morphology was evaluated using sagittal, axial, and coronal views. The measurements were conducted following the modified version of measurements described in the literature [27–29]. The detailed description of the landmarks and parameters used in the current study is given in Table 1.

For the sagittal view parameters, the slices that exhibited the greatest anteroposterior extent of the condylar head were selected. The sagittal view parameters are as follows (Figure 1):

- Anterior joint space (AJS);
- Superior joint space (SJS);
- Posterior joint space(PJS);
- Condyle head length (CHL);
- Condylar neck width (CNW).

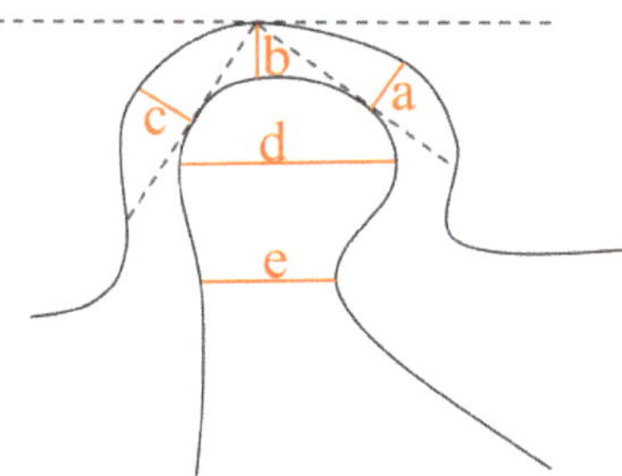

Figure 1. Parameters in sagittal view. a. AJS, b. SJS, c. PJS, d. CHL, e. CNW.

Table 1. Description of the landmarks and parameters.

	Name	Abbreviation	Description
General Landmarks	Sella	Se	The central point of the pituitary fossa of sphenoid bone
	Nasion	Na	The contact point between frontal bone, right nasal bone, and left nasal bone
	Basion	Ba	The central point on the anterior margin of the foramen magnum
	Menton	Me	Most inferior midpoint of the chin on the outline of the mandibular symphysis
	Midsagittal Plane	MSP	A plane passes through the points Se, Na, and Ba
	Superior Glenoid Fossa	SGF	The point that is placed at the most superior wall of glenoid fossa
	Anterior Tangent Line	ATL	A line connects SGF and the most prominent anterior aspect of the condyle
	Posterior Tangent Line	PTL	A line connects SGF and the most prominent posterior aspect of the condyle
Sagittal View	Anterior Condyle Point	ACP	The most anterior point of the condyle, which is situated on the anterior tangent line
	Superior Condyle Point	SCP	The most superior point of the condyle
	Posterior Condyle Point	PCP	The most posterior point of the condyle, which is situated on the posterior tangent line
	Anterior Joint Space	AJS	The perpendicular distance from ACP to glenoid fossa
	Superior Joint Space	SJS	The perpendicular distance from SCP to SGF
	Posterior Joint Space	PJS	The perpendicular distance from PCP to glenoid fossa
	Condyle Anterior Point	CAP	Most anterior point of the condyle corresponding to the area of maximum condyle length
	Condyle Posterior Point	CPP	Most posterior point of the condyle corresponding to the area of maximum condyle length
	Condyle Head Length	CHL	The distance between CAP and CPP
	Anterior Neck Point	ANP	Deepest point on the anterior aspect of condylar neck
	Posterior Neck Point	PNP	Deepest point on the posterior aspect of condylar neck
	Condylar Neck Width	CNW	Distance between ANP and CNP
Axial View	Minor Axis of the Condyle	MAC	The maximum diameter of condylar process in anteroposterior direction
	Long Axis of the Condyle	LAC	The maximum diameter of condylar process in mediolateral direction
	Condylar Axis Inclination	CI	The angle between midsagittal plane and the long axis of the condyle
Coronal View	Medial Condyle Point	MCP	Most medial point of the condylar head
	Lateral Condyle Point	LCP	Most lateral point of the condylar head
	Medial Joint Space	MJS	The shortest distance between MCP and medial wall of the glenoid fossa
	Lateral Joint Space	LJS	The shortest distance between the most LCP and lateral wall of the glenoid fossa

For axial view parameters, the slices that demonstrated the largest mesiodistal extent of the condylar head were selected in axial view. Axial view parameters are as follows (Figure 2):

- Minor axis of the condyle (MAC);
- Long axis of the condyle (LAC);
- Condylar axis inclination (CI).

For coronal view parameters, the slices that demonstrated the largest mesiodistal extent of each condylar head were selected in coronal view. The coronal view parameters and their descriptions are as follows (Figure 3):

- Medial joint space (MJS);
- Lateral joint space (LJS).

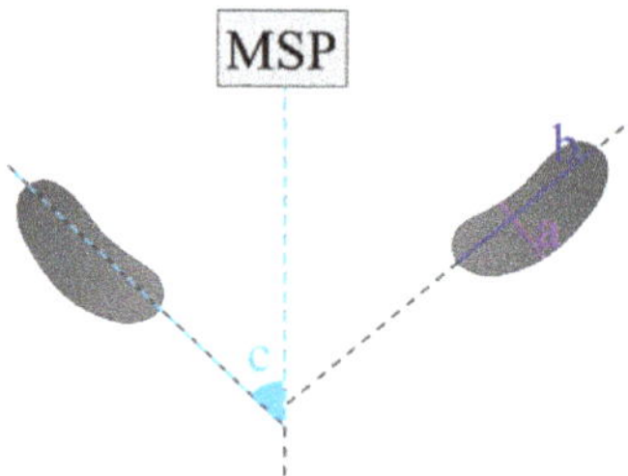

Figure 2. Parameters in axial view. a. MAC, b. LAC, c. CI.

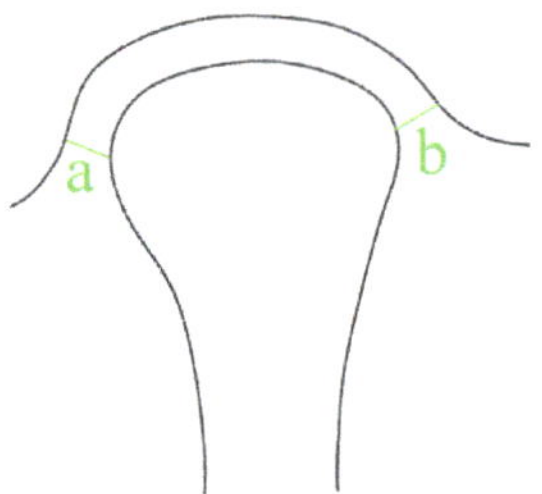

Figure 3. Parameters in coronal view. a. MJS, b. LJS.

Additional asymmetry index parameters were calculated for each parameter based on Habets' formula [30]. These parameters were utilized for the comparison between the study and control groups.

2.2. Statistical Analysis

The data were statistically analyzed using IBM SPSS Statistics Version 26.0 (Chicago, IL, USA). Pearson correlation analysis was performed to assess inter-examiner reliability. Confirmation of the data's normal distribution was evaluated using the Kolmogorov–Smirnov test due to the sample size exceeding 30.

Before conducting the paired samples *t*-test, the necessary assumptions were checked. The data were collected in pairs, and each observation within a pair was dependent on the other. The data exhibited a normal distribution, and the differences between paired observations showed equal variances. The values within each pair were not influenced by or related to the values in other pairs. As these assumptions were met, the paired samples *t*-test was performed.

The assumptions underlying Student's *t*-test were also checked. The normality assumption was met, and the variances were equal between the study and control groups. The observations within each group were independent of each other. Therefore, Student's *t*-test was performed.

In the first part of the study, the comparison of TMJ parameters between both sides of the face was analyzed using a paired samples *t*-test in the study group. Similarly, in the second part, the comparison of TMJ parameters between both sides of the face was also analyzed using a paired samples *t*-test in the control group. The lowest values of each parameter were recorded under "Side 1", while the highest values were recorded under "Side 2". For the analyses in the first and second parts, the lowest and highest values were used instead of distinguishing between right and left sides. In the third part of the study, Student's *t*-test was employed to compare the asymmetry indices of each TMJ parameter between the study group and the control group. A *p*-value of 0.05 was considered statistically significant.

3. Results

The correlation coefficients were 0.907 and 0.921 for the right side parameters and left side parameters, respectively (Table 2). These results indicated a high reliability between the measurements of the two examiners.

Table 2. Inter-examiner reliability.

	Intraclass Correlation	95% CI		F Test		
		Lower Bound	Upper Bound	F	df	*p*
Right Side	0.907	0.861	0.920	25.691	49	0.001 *
Left Side	0.921	0.874	0.952	30.678	49	0.001 *

* *p*-value is less than 0.05.

Furthermore, the overall data were found to be distributed normally based on the Kolmogorov–Smirnov test ($p > 0.05$). Consequently, parametric tests were applied in the present study. A paired samples *t*-test was conducted in the first and second parts of the study, while a Student's *t*-test was performed in the third part of the current study.

Among the total number of subjects, 40.2% were females and 59.8% were males. In the study group, there were 21 females (41.2%) and 30 males (58.8%), while the control group consisted of 20 females (39.2%) and 31 males (60.8%). The demographic characteristics of the subjects are presented in Table 3.

Table 3. Demographic characteristics of the subjects.

		N	%	Age Range	Mean	S.D.
	Female	21	41.2%	52–90	73.10	8.60
Study Group (PD)	Male	30	58.8%	51–88	72.41	10.82
	Total	51	100%	51–90	72.70	9.85
	Female	20	39.2%	51–91	72.30	10.63
Control Group	Male	31	60.8%	53–87	72.68	9.69
	Total	51	100%	51–91	72.52	9.96

Mean values of parameters for both TMJs in study and control groups are given in Table 4.

Table 4. Mean values of parameters for both TMJs in study and control groups.

	Study (*n* = 102)	Control (*n* = 102)
AJS	2.9 ± 1.3	3.4 ± 0.6
SJS	2.8 ± 1.3	3.5 ± 0.9
PJS	2.2 ± 1.0	2.3 ± 0.9
CHL	9.6 ± 1.8	10.0 ± 1.6
CNW	5.8 ± 1.4	6.0 ± 1.2
MAC	7.5 ± 1.6	8.0 ± 1.6
LAC	18.5 ± 2.5	18.8 ± 2.7
CI	68.67 ± 8.9	71.47 ± 7.14
MJS	2.7 ± 1.2	3.0 ± 0.7
LJS	2.6 ± 1.0	2.6 ± 0.7

The comparison of TMJ parameters between both sides of the face through paired samples *t*-test is given in Table 5 for the study group. All parameters that were measured on the three planes of the face showed a statistically significant difference between both sides of the face in the PD group ($p < 0.05$).

Table 5. Comparison of bilateral jaws in study group through paired samples *t*-test.

		Side 1	Side 2	p
Sagittal	AJS	2.5 ± 1.0	3.4 ± 1.4	0.001 *
	SJS	2.4 ± 1.0	3.2 ± 1.4	0.001 *
	PJS	1.9 ± 0.8	2.5 ± 1.0	0.001 *
	CHL	8.8 ± 1.7	10.4 ± 1.5	0.001 *
	CNW	5.2 ± 1.2	6.5 ± 1.2	0.001 *
Axial	MAC	6.9 ± 1.5	8.1 ± 1.5	0.001 *
	LAC	17.8 ± 2.4	19.2 ± 2.3	0.001 *
	CI	64.72 ± 7.68	72.61 ± 8.32	0.001 *
Coronal	MJS	2.2 ± 1.0	3.3 ± 1.2	0.001 *
	LJS	2.1 ± 0.7	3.1 ± 0.9	0.001 *

* *p*-value is less than 0.05.

The comparison of TMJ parameters between both sides of the face through paired samples *t*-test is given in Table 6 for the control group. AJS, SJS, PJS, CHL, CNW, MAC, CI, MJS, and LJS did not show a statistically significant difference between both sides of the face in control group. LAC showed a statistically significant difference between both sides of the face in control group ($p < 0.05$).

Table 6. Comparison of bilateral jaws in control group through paired samples *t*-test.

		Side 1	Side 2	p
Sagittal	AJS	3.2 ± 0.6	3.5 ± 0.6	0.100
	SJS	3.4 ± 0.9	3.6 ± 0.9	0.157
	PJS	2.3 ± 0.8	2.4 ± 0.9	0.243
	CHL	9.8 ± 1.6	10.2 ± 1.7	0.095
	CNW	5.9 ± 1.1	6.2 ± 1.1	0.210
Axial	MAC	7.8 ± 1.5	8.2 ± 1.7	0.190
	LAC	18.5 ± 2.9	19.0 ± 2.4	0.011 *
	CI	70.66 ± 6.96	72.29 ± 7.29	0.067
Coronal	MJS	2.9 ± 0.7	3.1 ± 0.6	0.254
	LJS	2.5 ± 0.7	2.7 ± 0.7	0.281

* *p*-value is less than 0.05.

Comparison of asymmetry indices between study and control groups through student *t*-test is given in Table 7. Index values of AJS, SJS, PJS, CHL, CNW, MAC, CI, MJS, and LJS showed a statistically significant difference between study and control groups ($p < 0.05$). LAC index did not show a statistically significant difference between study and control groups.

Table 7. Comparison of asymmetry indices between study and control groups through Student's *t*-test.

		Study ($n = 51$)	Control ($n = 51$)	p
Sagittal	AJS-Index	16.39 ± 10.62	3.76 ± 2.39	0.001 *
	SJS-Index	15.27 ± 12.15	3.97 ± 2.30	0.001 *
	PJS-Index	13.77 ± 10.30	3.67 ± 3.08	0.001 *
	CHL-Index	8.82 ± 7.86	1.99 ± 1.62	0.001 *
	CNW-Index	12.19 ± 11.44	2.52 ± 1.28	0.001 *
Axial	MAC-Index	8.36 ± 6.97	2.40 ± 5.97	0.001 *
	LAC-Index	3.91 ± 3.21	1.81 ± 6.68	0.064
	CI-Index	5.74 ± 3.87	1.13 ± 1.47	0.001 *
Coronal	MJS-Index	19.03 ± 14.28	3.95 ± 3.53	0.001 *
	LJS-Index	20.82 ± 14.94	4.65 ± 4.62	0.001 *

* *p*-value is less than 0.05.

4. Discussion

In the current study, a significant difference between the PD group and control group was observed with regard to TMJ morphology and symmetry.

The accurate imaging technique is crucial for assessment of the TMJ. Several imaging modalities have been utilized for evaluation of the TMJ area, including conventional radiography, computed tomography (CT), MR imaging, and cone-beam computed tomography (CBCT). Among these modalities, MR imaging offers exceptional quality images of soft tissues. It enables comprehensive evaluation of the soft tissue structures of the TMJ, synovial tissue, and articular disc. Therefore, the TMJ was evaluated through MR images in the present study.

In the first and second part of the current study, the comparisons were not performed based on the right and left sides of the face. The comparisons were performed based on the side that shows the highest values and the side that shows the lowest values of the parameters. Rather than focusing on right and left sides, the focus was on the difference between bilateral joints. In this way, the parameters that were effective in TMJ asymmetry were revealed through paired samples *t*-test. The statistical analyses that used in the current study allowed us to minimize personal differences between each subject. And, asymmetry indices for each parameter enabled us to compare study group and control group.

According to the results of the current study, healthy subjects showed morphological similarities and symmetry of the TMJ bilaterally in general (Figure 4). However, only the long axis of the condyle showed asymmetry in healthy subjects. On the other hand, patients with PD showed morphological differences and asymmetry of the TMJ bilaterally regarding all parameters measured in the three different planes (Figure 4). When the differences of both sides were compared between the healthy subjects and patients with PD, those with PD showed more asymmetrical TMJs in general except for the long axis of the condyle.

Although comparison of the parameters in PD group was impossible due to a lack of similar study design, the comparison of some parameters in control group was performed with previous studies that have control groups as well [27,29,31]. The mean value of AJS (3.4 ± 0.6) in the current study was higher compared to the previous studies (1.3 ± 0.2, 2.60 ± 0.79, 2.03 ± 0.50; and right: 2.1 ± 0.5; left: 1.9 ± 0.5) [27,31–33]. The higher values of AJS could be explained by the findings of a previous study that presents a tendency of increase in anterior space with age [34]. There was an important difference of sample age between the current study and above-mentioned previous studies. Therefore, lower values of AJS could be in accordance with the sample age between 12 and 59 years of age that was of previous studies' age range. The mean value of SJS (3.5 ± 0.9) in the current study was similar compared to previous studies (3.35 ± 0.73, 4.13 ± 1.03; and right: 3.2 ± 0.9; left: 3.4 ± 0.9) [27,32,33]. The mean value of PJS (2.3 ± 0.9) in the current study was similar compared to previous studies (2.1 ± 0.3, 2.49 ± 0.73, 2.64 ± 0.81; and right: 2.1 ± 0.7; left: 2.4 ± 0.8) [27,31–33]. The mean value of CHL (10.0 ± 1.6) in the current study was similar compared to a previous study conducted on TMJ radiography (right: 10.83 ± 1.65; left: 11.53 ± 2.82) [35].

The mean value of MAC (8.0 ± 1.6) in the current study was similar compared to a previous study (8.36 ± 1.29) [31]. The mean value of CI (71.47 ± 7.14) in the current study was similar compared to previous studies (71.88 ± 7.38, 75.46 ± 4.58) [31,32]. However, the mean value of CI in the current study was higher than another previous study (66.46 ± 8.91) [36]. The inconsistency between two studies could be explained by the records, which were both MR and CT images in the previous study. Moreover, the subjects were between 11 and 44 years of age, which was a range younger than the subjects of the current study. Changes that occur by aging could also be a possible reason for this difference.

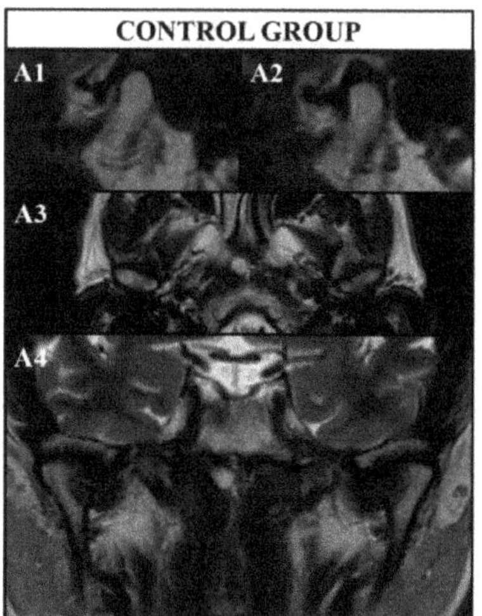
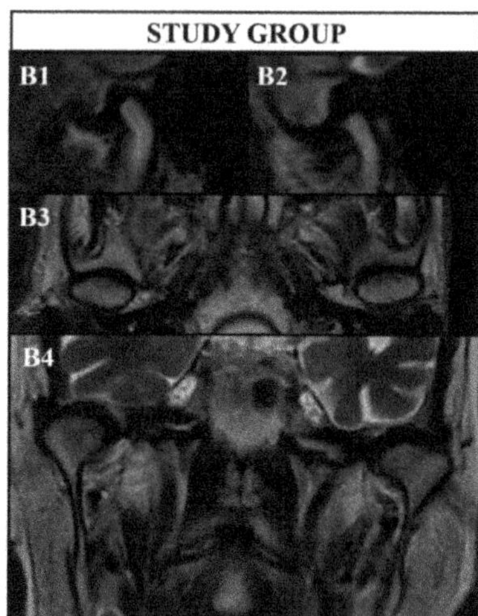

Figure 4. Images of bilateral TMJs from the study and control groups: (**A1,A2**) Healthy appearance of TMJ in sagittal view with symmetry of the right and left sides. (**A3**) Healthy appearance of TMJ in axial view with symmetry of the right and left sides. (**A4**) Healthy appearance of TMJ in coronal view with symmetry of the right and left sides. (**B1,B2**) Unhealthy appearance of TMJ in sagittal view with resorptive areas of the condyle head. Flattening of the condyle surface is observed in both TMJs, but it is more severe in (**B2**) compared to (**B1**) resulting in an asymmetrical appearance. An osteophyte is present at the anterior region of the condyle in (**B2**) further contributing to the asymmetry. (**B3**) Unhealthy appearance of TMJ in axial view with resorptive areas in each condylar head but in different localizations, resulting in an asymmetrical appearance of the right and left sides. (**B4**) Unhealthy appearance of TMJ in coronal view with asymmetry of right and left sides. An osteophyte is observed on the medial surface of the condyle on the right hand side. There is an increase in intensity of the inferior part of the lateral pterygoid muscle on left hand side, which may be related to myospasms.

The mean value of MJS (3.0 ± 0.7) in the current study was similar compared to previous studies (2.54 ± 0.80, 2.75 ± 0.90; and right: 2.9 ± 1; left: 2.9 ± 0.9) [31–33]. The mean value of LJS (2.6 ± 0.7) in the current study was similar compared to a previous study (right: 2.6 ± 0.9; left: 2.4 ± 0.8) [33].

The association between PD and TMDs was assessed in previous studies with different methodologies. According to a previous survey study, 110 patients with PD were underwent comprehensive evaluation to show the frequency of TMD symptoms in patients with PD. It was revealed that the prevalence of TMDs was found to be 35%, exhibiting a higher occurrence among males (58%) and elderly individuals (53%) [24]. Higher occurrence of TMDs among patients diagnosed with PD was also presented in another previous cross-sectional study [23]. In a previous cohort study, it was reported that a 2.11-fold increase in TMD risk exists for individuals with PD compared to healthy people [25]. Another previous cohort study that involves a total of 514.866 participants consisted of two parts. According to Part I, a cohort of 4.455 individuals diagnosed with TMD was meticulously matched with 17.820 control participants, maintaining a ratio of 1:4. Similarly, in Part II, a group of 6.076 individuals with PD was carefully matched with 24.304 control participants, also at a ratio of 1:4 [37].

Although the aforementioned studies have revealed an association between PD and TMD, there is still a debate as to whether TMDs are a risk factor for PD, or whether PD is a risk factor for TMDs. In first scenario, considering the important neural network in the TMJ area, the improvement of PD symptoms after intervention in the TMJ may affirm that TMDs can be a risk factor for PD. Neural connections contribute to the sensory and motor functions of the TMJ, allowing for sensory perception, proper jaw movement, and control of the related muscles. The TMJ is innervated by branches of the trigeminal nerve (fifth cranial nerve), specifically its mandibular branch. This nerve provides sensory innervation to the TMJ, transmitting information related to pain, touch, and temperature. The sensory signals originating from the TMJ is transmitted to the trigeminal sensory nuclei situated in the brainstem. Within these nuclei, the sensory information is processed and then transmitted to higher brain regions that are responsible for tasks such as pain perception and motor control. The sensory input originating from the TMJ is further relayed to different regions of the central nervous system, including the thalamus and somatosensory cortex. This transmission enables the perception and interpretation of sensations arising from the TMJ, allowing for the brain to process and make sense of the sensory information received from the TMJ region. The motor control of the muscles related to the TMJ, such as the masticatory muscles, relies on the activation of motor neurons located in the brainstem. These motor neurons are responsible for transmitting signals to the muscles, thereby facilitating the movements of the jaw involved in activities such as chewing, speaking, and other functions. The TMJ is intricately connected to the autonomic nervous system, which governs various involuntary processes in the body. Autonomic fibers, comprising both sympathetic and parasympathetic fibers, extend their influence to the TMJ region. These fibers have the capacity to impact factors, such as blood flow, inflammation, and other physiological responses within the TMJ area. A previous case report showed improvement in motor dysfunction of PD after usage of bite splint together with oral medication [38]. Although the first scenario needs to be supported by new studies, the second scenario is a more widely accepted approach considering the symptoms of PD. The symptoms of PD including rigidity, muscular stiffness, and tremor of the masticatory muscles can lead to clenching, bruxism, and TMDs [18,19,39]. Moreover, considering that Parkinson's disease primarily impacts the motor system, it is probable that the consequent motor symptoms directly contribute to the occurrence of TMDs.

Although it is impossible to imply whether TMJ asymmetry is a risk factor for PD or whether PD is a risk factor for TMJ asymmetry, the current study presents an association between PD and TMJ asymmetry. Based on the study's results, preventive measures can be implemented clinically to address the risk factors for both TMDs and PD. Patients with PD can be made aware of the benefits of physical therapy for jaw exercises, making changes to their daily life habits and dietary habits, and managing bruxism. These interventions can contribute to improving patients' overall well-being. Bruxism has been observed to alter activity in subcortical, cortical, and spinal circuits, as well as modify deep tendon reflexes in the extremities [40,41]. Previous studies have revealed that muscular tone in various regions of the body can be influenced by mechanosensory signals originating from the jaws, transmitted through the trigeminal nerve [40,42]. Therefore, clenching, bruxism, asymmetry of the TMJs, and morphological differences in the TMJs may be considered as potential risk factors associated with future PD. Consequently, regular examinations of the TMJs can be implemented as part of routine healthcare practices for healthy individuals.

Condylar position can vary among individuals with different skeletal patterns [43]. However, in the current study, the skeletal patterns of the subjects could not be considered due to the limited number of subjects with specific diseases who also met the selection criteria. Therefore, future studies could establish selection criteria based on skeletal patterns or subgroups classified according to different skeletal patterns. Hence, the measurements could be more sensitive by eliminating the influence of different condylar position in different skeletal patterns.

Previously acquired brain MR images were used in the current retrospective study. The primary purpose of obtaining these images was for neurological evaluation, not TMJ assessment. Therefore, the position of the mouth (open or closed) was not taken into consideration when acquiring these images. Although efforts were made to mitigate the potential impact of this factor by selecting the paired samples *t*-test and using of index values, in future prospective studies, it may be considered to acquire images with both an open and closed mouth to address this limitation.

The other limitations of present study were the confounding factors, which may also affect the morphology and symmetry of TMJ. These factors could not be considered during selection the subjects due to the lack of previous records and limited sample size of subjects with a specific disease. Factors such as dental history, TMD symptoms, current dental situation, craniofacial pattern, and malocclusion could be other factors that may affect the morphology and asymmetry of the TMJ, thereby influencing the results of the current study. Rather than using brain MR images, the MR images obtained specifically from the TMJ region would provide more accurate measurements. The records utilized in the present study were readily available for neurological assessments. However, when considering the relationship between PD and the TMJ, the inclusion of TMJ images has the potential to be incorporated into the standard initial records for Parkinson's patients in the future.

To the best of our knowledge this was the first study that investigates the association between the TMJ and PD on a morphological basis. Current results could offer new study designs in order to assess the role of TMJ morphology in development of PD with a larger sample size and with less confounding factors.

5. Conclusions

The null hypotheses were partly rejected. Patients with PD showed asymmetry in their bilateral temporomandibular joints regarding to joint spaces in sagittal view, condyle head length, condyle neck width, the long axis of the condyle, the minor axis of the condyle, condylar inclination, and joint spaces in sagittal view. Healthy individuals demonstrated symmetry in their bilateral temporomandibular joints for all these parameters except for the long axis of the condyle. There was difference in temporomandibular joint symmetry between healthy individuals and those with PD in terms of joint spaces in sagittal view, condyle head length, condyle neck width, the minor axis of the condyle, condylar inclination, and joint spaces in sagittal view.

Author Contributions: Conceptualization, A.S.A.; methodology, A.S.A.; software, A.H.A.K.; validation, A.H.A.K.; formal analysis, A.H.A.K.; investigation, A.S.A.; resources, A.H.A.K.; data curation, A.S.A. and A.H.A.K.; writing—original draft preparation, A.S.A.; writing—review and editing, A.S.A.; visualization, A.H.A.K.; supervision, A.S.A.; project administration, A.S.A. and A.H.A.K. All authors have read and agreed to the published version of the manuscript.

Funding: This research received no external funding.

Institutional Review Board Statement: The study was conducted in accordance with the Declaration of Helsinki and approved by the Ethics Committee of NECMETTIN ERBAKAN UNIVERSITY (protocol code: 2023/4325, date of approval: 5 May 2023).

Informed Consent Statement: Patient consent was waived due to the retrospective design of this archive study, which only use radiographic images of the patients with no intervention to patients.

Data Availability Statement: Research data are available upon request due to privacy or ethical restrictions.

Conflicts of Interest: The authors declare no conflict of interest.

References

1. Cooper, B.C.; Kleinberg, I. Examination of a large patient population for the presence of symptoms and signs of temporomandibular disorders. *Cranio* **2007**, *25*, 114–126. [CrossRef] [PubMed]
2. Gauer, R.L.; Semidey, M.J. Diagnosis and treatment of temporomandibular disorders. *Am. Fam. Physician* **2015**, *91*, 378–386. [PubMed]
3. Valesan, L.F.; Da-Cas, C.D.; Réus, J.C.; Denardin, A.C.S.; Garanhani, R.R.; Bonotto, D.; de Souza, B.D.M. Prevalence of temporomandibular joint disorders: A systematic review and meta-analysis. *Clin. Oral Investig.* **2021**, *25*, 441–453. [CrossRef]
4. Chisnoiu, A.M.; Picos, A.M.; Popa, S.; Chisnoiu, P.D.; Lascu, L.; Picos, A.; Chisnoiu, R. Factors involved in the etiology of temporomandibular disorders—A literature review. *Clujul. Med.* **2015**, *88*, 473–478. [CrossRef] [PubMed]
5. GBD 2016 Neurology Collaborators. Global, regional, and national burden of neurological disorders, 1990–2016: A systematic analysis for the Global Burden of Disease Study 2016. *Lancet Neurol.* **2019**, *18*, 459–480. [CrossRef]

6. Kalia, L.V.; Lang, A.E. Parkinson's disease. *Lancet* **2015**, *386*, 896–912. [CrossRef] [PubMed]
7. Salat, D.; Tolosa, E. Levodopa in the treatment of Parkinson's disease: Current status and new developments. *J. Park. Dis.* **2013**, *3*, 255–269. [CrossRef]
8. Simon, D.K.; Tanner, C.M.; Brundin, P. Parkinson disease epidemiology, pathology, genetics, and pathophysiology. *Clin. Geriatr. Med.* **2020**, *36*, 1–12. [CrossRef]
9. Lesage, S.; Brice, A. Parkinson's disease: From monogenic forms to genetic susceptibility factors. *Hum. Mol. Genet.* **2009**, *18*, R48–R59. [CrossRef]
10. Hirsch, E.C.; Hunot, S. Neuroinflammation in Parkinson's disease: A target for neuroprotection? *Lancet Neurol.* **2009**, *8*, 382–397. [CrossRef]
11. Schapira, A.H.V.; Jenner, P. Etiology and pathogenesis of Parkinson's disease. *Mov. Disord.* **2011**, *26*, 1049–1055. [CrossRef] [PubMed]
12. Chaudhuri, K.R.; Odin, P.; Antonini, A.; Martinez-Martin, P. Parkinson's disease: The non-motor issues. *Park. Relat. Disord.* **2011**, *17*, 717–723. [CrossRef]
13. Postuma, R.B.; Berg, D.; Stern, M.; Poewe, W.; Olanow, C.W.; Oertel, W.; Obeso, J.; Marek, K.; Litvan, I.; Lang, A.E.; et al. MDS clinical diagnostic criteria for Parkinson's disease. *Mov. Disord.* **2015**, *30*, 1591–1601. [CrossRef] [PubMed]
14. Broen, M.P.; Narayen, N.E.; Kuijf, M.L.; Dissanayaka, N.N.; Leentjens, A.F. Prevalence of anxiety in Parkinson's disease: A systematic review and meta-analysis. *Mov. Disord.* **2016**, *31*, 1125–1133. [CrossRef] [PubMed]
15. Baiano, C.; Barone, P.; Trojano, L.; Santangelo, G. Prevalence and clinical aspects of mild cognitive impairment in Parkinson's disease: A meta-analysis. *Mov. Disord.* **2020**, *35*, 45–54. [CrossRef]
16. Chen, H.; Zhao, E.J.; Zhang, W.; Lu, Y.; Liu, R.; Huang, X.; Ciesielski-Jones, A.J.; Justice, M.A.; Cousins, D.S.; Peddada, S. Meta-analyses on prevalence of selected Parkinson's nonmotor symptoms before and after diagnosis. *Transl. Neurodegener.* **2015**, *4*, 1–8. [CrossRef] [PubMed]
17. Siciliano, M.; Trojano, L.; Santangelo, G.; De Micco, R.; Tedeschi, G.; Tessitore, A. Fatigue in Parkinson's disease: A systematic review and meta-analysis. *Mov. Disord.* **2018**, *33*, 1712–1723. [CrossRef]
18. Laine, C.M.; Yavuz, S.U.; D'Amico, J.M.; Gorassini, M.A.; Turker, K.S.; Farina, D. Jaw tremor as a physiological biomarker of bruxism. *Clin. Neurophysiol.* **2015**, *126*, 1746–1753. [CrossRef]
19. Verhoeff, M.C.; Lobbezoo, F.; Wetselaar, P.; Aarab, G.; Koutris, M. Parkinson's disease, temporomandibular disorders and bruxism: A pilot study. *J. Oral Rehabil.* **2018**, *45*, 854–863. [CrossRef]
20. Ross, C.A.; Smith, W.W. Gene–environment interactions in Parkinson's disease. *Park. Relat. Disord.* **2007**, *13*, S309–S315. [CrossRef]
21. Deogaonkar, M.; Patel, B.C.; Ghosh, D.; Kukreja, R.; Subramaniam, K. *The Neuroanatomy, Neurophysiology, and Clinical Pharmacology of Motor Neurons and Motor Pathways*; StatPearls Publishing: St. Petersburg, FL, USA, 2020.
22. Silva, T.V.D.A.; Coriolano, M.D.G.W.D.S.; Lins, C.C.D.S.A. Temporomandibular joint dysfunction in Parkinson's disease: An integrative literature review. *Rev. CEFAC.* **2017**, *19*, 702–711. [CrossRef]
23. Baumann, P.; Sági, A.; Perjés, B.; Prémusz, V.; Ács, P.; Marada, G.; Kovács, N.; Radnai, M. Temporomandibular joint disorder in patients with Parkinson's disease—a pilot study. *Health Probl. Civil.* **2020**, *14*, 235–241. [CrossRef]
24. Silva, R.M.D.; Santos, V.L.D.; Silva, T.V.D.A.; Lins, C.C.D.S.A. Prevalence of temporomandibular joint disorder in people with Parkinson's disease in a public university hospital. *Rev. CEFAC* **2019**, *21*, e17618. [CrossRef]
25. Chen, Y.Y.; Fan, H.C.; Tung, M.C.; Chang, Y.K. The association between Parkinson's disease and temporomandibular disorder. *PLoS ONE* **2019**, *14*, e0217763. [CrossRef]
26. da Costa Silva, P.F.; Biasotto-Gonzalez, D.A.; Motta, L.J.; Silva, S.M.; Ferrari, R.A.M.; Fernandes, K.P.S.; Bussadori, S.K. Impact in oral health and the prevalence of temporomandibular disorder in individuals with Parkinson's disease. *J. Phys. Ther. Sci.* **2015**, *27*, 887–891. [CrossRef] [PubMed]
27. Ikeda, K.; Kawamura, A. Assessment of optimal condylar position with limited cone-beam computed tomography. *Am. J. Orthod. Dentofac. Orthop.* **2009**, *135*, 495–501. [CrossRef] [PubMed]
28. Rodrigues, A.F.; Fraga, M.R.; Vitral, R.W.F. Computed Tomography Evaluation of the Temporomandibular Joint in Class II Division 1 and Class III Malocclusion Patients: Condylar Symmetry and Condyle-Fossa Relationship. *Am. J. Orthod. Dentofac. Orthop.* **2009**, *136*, 199–206. [CrossRef]
29. Al-koshab, M.; Nambiar, P.; John, J. Assessment of condyle and glenoid fossa morphology using CBCT in South-East Asians. *PLoS ONE* **2015**, *10*, e0121682. [CrossRef]
30. Habets, L.L.M.H.; Bezuur, J.N.; Naeiji, M.; Hansson, T.L. The Orthopantomogram®, an aid in diagnosis of temporomandibular joint problems. II. The vertical symmetry. *J. Oral Rehabil.* **1988**, *15*, 465–471. [CrossRef]
31. -Abdulqader, A.A.; Ren, L.; Alhammadi, M.; Abdu, Z.A.; Mohamed, A.A.S. Three-dimensional analysis of temporomandibular joint in Chinese adults with normal occlusion and harmonious skeleton. *Oral Radiol.* **2020**, *36*, 371–382. [CrossRef]
32. Alhammadi, M.S.; Shafey, A.S.; Fayed, M.S.; Mostafa, Y.A. Temporomandibular joint measurements in normal occlusion: A three-dimensional cone beam computed tomography analysis. *J. World Fed. Orthod.* **2014**, *3*, 155–162. [CrossRef]
33. Dalili, Z.; Khaki, N.; Kia, S.J.; Salamat, F. Assessing joint space and condylar position in the people with normal function of temporomandibular joint with cone-beam computed tomography. *Dent. Res. J.* **2012**, *9*, 607. [CrossRef]
34. Li, Y.; Zhou, W.; Wu, Y.; Dai, H.; Zhou, J. The relation between incisal guidance angle and the growth and development of temporomandibular joint: A multi-cross-sectional retrospective study. *BMC Oral Health* **2021**, *21*, 1–10. [CrossRef] [PubMed]

35. Manja, C.D.; Rajaduray, D. Analysis of height and width of mandibular condyle and shape of the articular eminence with and without clicking using tmj radiography. *Int. J. Dent. Res.* **2019**, *4*, 99–103. [CrossRef]
36. De Stefano, A.A.; Guercio-Monaco, E.; Hernández-Andara, A.; Galluccio, G. Association between temporomandibular joint disc position evaluated by magnetic resonance imaging and mandibular condyle inclination evaluated by computed tomography. *J. Oral Rehabil.* **2020**, *47*, 743–749. [CrossRef]
37. Choi, H.-G.; Yoon, J.-H.; Chung, T.-H.; Min, C.; Yoo, D.-M.; Wee, J.-H.; Kang, S.-Y.; Choi, Y.; Hong, S.-J.; Byun, S.-H. Association between Temporomandibular Joint Disorder and Parkinson's Disease. *Brain Sci.* **2021**, *11*, 747. [CrossRef]
38. Nomoto, S.; Nakamura, M.; Sato, T.; Hisanaga, R. Occlusal treatment with bite splint improves dyskinesia in Parkinson's disease patient: A case report. *Bull. Tokyo Dent. Coll.* **2013**, *54*, 157–161. [CrossRef] [PubMed]
39. Verhoeff, M.C.; Koutris, M.; van Selms, M.K.A.; Brandwijk, A.N.; Heres, M.S.; Berendse, H.W.; van Dijk, K.D.; Lobbezoo, F. Is dopaminergic medication dose associated with self-reported bruxism in Parkinson's disease? A cross-sectional, questionnaire-based study. *Clin. Oral Investig.* **2021**, *25*, 2545–2553. [CrossRef]
40. Takahashi, M.; Ni, Z.; Yamashita, T.; Liang, N.; Sugawara, K.; Yahagi, S.; Kasai, T. Excitability changes in human hand motor area induced by voluntary teeth clenching are dependent on muscle properties. *Exp. Brain Res.* **2006**, *171*, 272–277. [CrossRef] [PubMed]
41. Faulkner, K.D. Bruxism: A review of the literature. Part I. *Aust. Dent. J.* **1990**, *35*, 266–276. [CrossRef]
42. Demerjian, G.G.; Barkhordarian, A.; Chiappelli, F. Testing patient targeted therapies in patients with temporomandibular joint disorder with the arthrokinetic reflex: Individual patient research. *J. Transl. Med.* **2016**, *14*, 1–5. [CrossRef] [PubMed]
43. Fraga, M.R.; Rodrigues, A.F.; Ribeiro, L.C.; da Silva Campos, M.J.; Vitral, R.W.F. Anteroposterior condylar position: A comparative study between subjects with normal occlusion and patients with Class I, Class II Division 1, and Class III malocclusions. *Med. Sci. Monit.* **2013**, *19*, 903. [CrossRef] [PubMed]

Review

Diagnosis and Treatment of Myogenous Temporomandibular Disorders: A Clinical Update

Natalie Hoi Ying Chan [1], Ching Kiu Ip [1], Dion Tik Shun Li [2] and Yiu Yan Leung [2,*]

1   Faculty of Dentistry, The University of Hong Kong, Hong Kong
2   Oral and Maxillofacial Surgery, Faculty of Dentistry, The University of Hong Kong, Hong Kong
*   Correspondence: mikeyyleung@hku.hk

Abstract: Myogenous temporomandibular disorders (M-TMDs) are the most common chronic orofacial pain, affecting the masticatory muscles and, thus, jaw movement. While a concise diagnosis is crucial to formulate a rational treatment plan, the similarities in clinical presentations that M-TMDs share with other neuromuscular disorders affecting the temporomandibular joint (TMJ) could easily confuse physicians. In addition to the basics, such as thorough history taking and meticulous clinical examinations, different imaging techniques are useful adjuncts to facilitate the diagnostic process. This review presents an overview of the current understanding on a variety of diagnostic and treatment modalities for M-TMD patients. It is essential to highlight that there is not a single treatment for all, and the benefits of multidisciplinary strategies have been noted for the effective management of myogenous TMD pain. Treatment modalities ranging from conservative to minimally invasive options are discussed in this review.

Keywords: temporomandibular disorders; temporomandibular joint dysfunction syndrome; facial pain; temporomandibular joint; myalgia

check for **updates**

Citation: Chan, N.H.Y.; Ip, C.K.; Li, D.T.S.; Leung, Y.Y. Diagnosis and Treatment of Myogenous Temporomandibular Disorders: A Clinical Update. *Diagnostics* **2022**, *12*, 2914. https://doi.org/10.3390/diagnostics12122914

Academic Editor: Luis Eduardo Almeida

Received: 24 October 2022
Accepted: 19 November 2022
Published: 23 November 2022

Publisher's Note: MDPI stays neutral with regard to jurisdictional claims in published maps and institutional affiliations.

1. Introduction

Temporomandibular disorders (TMDs) refer to a heterogeneous group of musculoskeletal dysfunctions affecting the temporomandibular joint (TMJ) and/or the masticatory muscles [1] that control jaw movement. According to Diagnostic Criteria for TMD (DC/TMD) Axis I, TMDs are divided into Group I: muscle disorders (including myofascial pain with and without mouth-opening limitation; Group II: involving disc displacement with or without reduction and mouth-opening limitation; and Group III: arthralgia, arthritis, and arthrosis [2]. They are the most common chronic orofacial pain, affecting 31% of adults and 11% of children and adolescents among the general population [3,4] TMDs and myogenous temporomandibular disorders (M-TMDs), also known as masticatory myalgia, are the most common condition, affecting 45.3% of TMD cases [2,5–8].

While there has been extensive research on TMD, the pathophysiology is not completely understood. TMDs have a multifactorial aetiology, and among them, some researchers reported that central sensitisation may play a role in chronic pain in M-TMD patients. Contrary to arthrogenous TMD, which appears to be a localized phenomenon, myogenous TMD may present overlapping features with other disorders, such as fibromyalgia and primary headaches, characterized by chronic primary pain related to dysfunction of the central nervous system (CNS), probably through the phenomenon of central sensitisation. Thus, central sensitisation could represent the basis of chronic pain, "or pain that persists beyond a normal time of healing" in patients affected by TMD [9]. However, TMDs of myogenous origin are generally considered to be strongly associated with psychogenic factors such as psychological stress, anxiety, depression, sleep and hormonal disturbances [10]. Some researchers reported that patients diagnosed with myofascial pain have more severe depressive and nonspecific physical symptoms than patients diagnosed with TMJ internal derangement (i.e., disk displacement) [11,12]. Factors like facial asymmetry and other forms of dentofacial deformities are proven to be correlated with TMD because of imbalanced masticatory forces, while the correction of these deformities could bring improvements to the TMD symptoms [13–15]. Recent studies have also suggested that during periods of ongoing lockdown and isolation due to the COVID-19 pandemic, people who were frequently under stress and anxiety may be more likely to develop depression and TMD symptoms [16]. One theory holds that these people may be more likely to clench their muscles [8], a form of repeated strain that leaves muscles susceptible to myofascial trigger points (MTrPs) [17,18]. Moreover, patients with more severe signs and symptoms of TMD had a lower pressure pain threshold [19]. These findings that psychological variables are closely tied to the development of TMD have been confirmed by the Orofacial Pain: Prospective Evaluation and Risk Assessment (OPPERA) study [20].

Intriguingly, sleep bruxism was also found to have a positive correlation with myofascial pain, arthralgia and joint pathology, such as disc displacement and joint noises. In articles that used polysomnographic analysis (PSG) for bruxism diagnosis, a positive correlation was established between bruxism and masticatory muscular activity of the temporal and masseter muscles during sleep. It is demonstrated that the increase in EMG activity during sleep could be a risk factor for myofascial pain. Thus, it is possible to suggest that bruxism would be associated with TMD [21].

Despite the research effort on M-TMD up until now, establishing a correct diagnosis which is critical for the selection of the proper treatment remains a challenge for physicians. This is because the presentation of M-TMD may mimic other non-TMD conditions and requires a more comprehensive patient assessment. At present, there is no simple one-step diagnostic test to help pinpoint a definitive diagnosis of M-TMD.

This article aims to provide an overview of the current diagnostic and treatment modalities available in the management of M-TMD. Both conservative and minimally invasive options will be discussed, as there is not a single treatment for M-TMD which encompasses a wide range of diagnoses.

2. Diagnoses and Classifications

According to the new evidence-based Diagnostic Criteria for TMD (Axis I DC/TMD protocol) [2], muscle pain myalgia represents what was called myofascial pain in RDC/TMD. The term myofascial pain now describes two new DC/TMD diagnoses: myofascial pain and myofascial pain with referral.

For the new DC/TMD classification, myogenous TMD pain diagnoses are organized into four major subclasses: myalgia, tendonitis, myositis and spasm (Table 1). Myalgia is further subdivided into local myalgia, myofascial pain and myofascial pain with referral upon myofascial examination protocol (Table 2).

Myalgia is described as pain of muscle origin that is affected by jaw movement, function or parafunction, and replication of this pain occurs with provocation testing of the masticatory muscles. Patients with TMD will experience pain in the jaw, temple, ear or front of ear and pain modified with jaw movement, function or parafunction. This is acute to a chronic condition that includes the presence of regional pain associated with tender areas referred to as trigger points (TrPs), which are expressed in taut bands of skeletal muscles, tendons or ligaments [22]. Although the pain occurs most often in the region over the TrPs, pain can be referred to areas distant from the TrPs (e.g., temporalis, referring to the frontal area, and masseter, referring to the ear or the posterior teeth). Often, reproducible duplication of pain complaints with specific palpation of the tender area is diagnostic.

Table 1. Differential diagnosis of M-TMDs according to the DC/TMD classification.

Differential Diagnosis of M-TMD

1. Muscle pain
    A. Myalgia
    B. Tendonitis
    C. Myositis
    D. Spasm

2. Contracture
3. Hypertrophy
4. Neoplasm

5. Movement Disorders
    A. Orofacial dyskinesia
    B. Oromandibular dystonia

6. Masticatory muscle pain related to central/systemic pain disorder   A. Fibromyalgia/widespread pain

Table 2. Subdivision of myalgia according to the DC/TMD classification.

Classification of M-TMD	Clinical Findings
Myalgia	Familiar pain in the masseter and temporalis upon palpation or mouth opening
Local myalgia	Familiar pain in the masseter and temporalis localized to the site of palpation
Myofascial pain	Pain in the masseter and temporalis spreading beyond the site of palpation but within the confines of the muscle being palpated
Myofascial pain with referral	Pain in the masseter and temporalis beyond the confines of the muscle being palpated

The diagnostic algorithms in the new DC/TMD for myalgia now include criteria for modification of pain by function, movement or parafunction; these criteria are also included in the TMD Pain Screener [23]. Currently, the clinical examination for myalgia includes pain with jaw-opening movements and palpation of the temporalis and masseter muscles. Pain from these provocation tests must replicate the patient's pain complaint. To differentiate the three types of myalgia, the duration of the 1 kg of palpation pressure is increased to 5 s to allow more time to elicit spreading or referred pain, if present. Pain is localized during palpation (local myalgia) or spreads within (myofascial pain) or beyond (myofascial pain with referral) the palpated muscular territory. If a diagnosis of myalgia is desired and no distinction between the three types is needed, the disorder of myofascial pain with a limited opening, as described in the RDC/ TMD, is eliminated.

Although tendonitis, myositis and spasm were less routinely encountered, it is important to include them in the differential diagnosis of TMJ disorders and pains. In addition, problems such as neoplasms, migraine, neuralgia and psychological disorders should also be considered. One case-control study [24] found that the diagnosis of myofascial pain is significantly higher in fibromyalgia patients. According to this line of thinking, one must not forget that TMD symptoms, which are difficult to diagnose and often missed, should be kept in mind in the management of fibromyalgia patients.

3. Diagnostic Approaches

Given the multifactorial aetiological nature of TMDs, a thorough history and clinical examination remain the cornerstones for the diagnosis of TMD [8].

Assessment of TMD patients should begin with a history taking of pain which follows the same format for other pain conditions [25]. Common chief complaints include pain on chewing/mouth opening, headache, ear pain, lack of chewing force, easily induced fatiguability in the masticatory muscles and disability to open the mouth wide [26]. Patients with myogenous TMD usually describe the condition to be a spontaneous dull aching pain and localized tenderness or stiffness in the masticatory muscles. A visual analogue scale (VAS) should be used to grade the severity of pain, so treatment progress can be quantitatively monitored. A past and current medical history, including a full medications list, may reveal any comorbidities that may be related to TMD. The clinician should also pay attention to any habits such as smoking, drinking and recreational drug use, and any history of clenching or bruxism as reported in complaints by the patient's bed partner. Additionally, the clinician should ask questions regarding stress levels or the presence of psychiatric illnesses such as anxiety and depression, as they are consistently associated with TMD. Although most clinicians treating TMD are experienced in obtaining a clinical history, some may not be comfortable with taking a psychological history. If desired, the clinician may employ numerous psychosocial instruments available to aid in their diagnosis, such as those in Axis-II of DC/TMD [2]. When necessary, the patient may be referred for a psychological evaluation.

Diagnosis of myogenous TMD relies largely on physical palpation of the soft tissue by a trained physician [27] and the recognition of pain points by the patient. The confirmation of the location of pain in the masseter and temporalis muscle, and whether the pain is confined, remains within or spreads beyond the confines of the muscle, should be confirmed with a muscle and MTrP palpation, which is carried out at rest and during mandibular function. A steady firm pressure (~2 kg) should be applied firmly over the muscle of concern when in the relaxed state for at least 1–2 s, and the patient should be asked to rate the tenderness during the palpation. Palpation should follow the direction of the muscle fibres to detect taut bands and check for the presence of MTrPs which produce referred pain to a nearby site.

Maximum mouth opening (MMO), including pain-free maximum mouth opening, maximum unassisted mouth opening and maximum assisted mouth opening, should be assessed in each appointment [28]. Any pain with maximum unassisted or assisted opening should also be noted.

3.1. Imaging Modalities

Ideally, the diagnosis of TMD pain is reached by a combination of clinical manifestations and diagnostic imaging confirmation. When further imaging is desired, CBCT and MRI are the diagnostic imaging techniques most commonly used in the field of dentistry to aid the diagnosis of TMD [29]. While CBCT is optimal for viewing skeletal and dental tissues which are especially useful in identifying degenerative joint diseases such as osteoarthritis [30], MRI is considered to be the gold standard when assessing the articular disc in terms of location and morphology [29,31] as MRI can provide early detection of disc abnormalities and the presence of joint effusion [32]. All disc-related TMD problems can be confirmed by MRI when indicated [31].

Regarding TMD with myogenous origin, whereas MTrPs are one of the most common and important causes of musculoskeletal pain, detection of MTrPs is critical for more comprehensive clinical evaluation and treatment of TMD patients. Unfortunately, CBCT and MRI are not able to detect the presence of taut bands or MTrPs. Nevertheless, when further information is desired, for example, the correct localization of MTrPs needed for different pain relief techniques, notably dry needling and injection-based therapies, ultrasonography may be a viable tool for locating the taut band or MTrPs [33].

Diagnostic ultrasound (US) has been proposed as a method to improve the reliability of MTrPs' localization as it is relatively cheap and accessible. Although the diagnostic efficacy of US is highly dependent on the operator's skills, knowledge of compartmental muscle anatomy and experience in assessing normal and abnormal muscle tissue in the static and dynamic state [34], many articles have shown that US could identify MTrP on upper trapezius muscle or other musculature [33,35–37]. Yet, little evidence was found for muscles of mastication. Our centre is now conducting a clinical trial on patients presented with myogenous TMD to assess the diagnostic efficacy of US, and its findings will be presented in due course.

3.2. Diagnosis of M-TMD

The International Association for the Study of Pain Subcommittee on Taxonomy [38] has classified myofascial pain as pain in any muscle with MTrPs that are very painful upon compression during palpation and cause referred pain.

Myofascial pain is diagnosed in five scenarios according to Shah et al. [27]: (1) when the clinician feels a hyperirritable spot within a palpably taut band of muscle fascia; (2) upon sustained compression of this hyperirritable spot, the patient reports new or increased dull aching pain in a nearby site; (3) when a decreased range of unassisted movement of the involved body area is noted; (4) weakness without atrophy and no neurological deficit explaining this weakness; (5) the presence of referred autonomic phenomena upon compression of the hyperirritable spot and/or a twitch response to snapping palpation of the taut bands.

4. Treatment Modalities

The treatment approach of TMD can be broadly classified into three types: conservative, minimally invasive and invasive [39]. Currently, the paradigm has shifted from open procedures to non-invasive options [8,40]. Thus, reversible conservatory approaches are usually considered the first line of treatment [8,41]. As surgical approaches usually target arthrogenous TMD, they are not the focus of this article. Several treatment modalities have been reported to successfully treat M-TMD by pain relief and restoring mouth opening. Nonetheless, to date, there is still a lack of consensus and evidence as to which is the most preferred option [42–45].

4.1. Conservative Therapies

4.1.1. Pharmacological Therapy

Medications are often prescribed initially as a non-invasive measure to treat TMD [8,46]. The most commonly used pharmacological agents are non-steroidal anti-inflammatory drugs (NSAIDs), muscles relaxants and anti-depressants. Despite carrying some well-known adverse effects, especially gastrointestinal disturbance, the anti-inflammatory and analgesics properties of NSAIDs, for example, ibuprofen, naproxen, diclofenac etc., render them popular among clinicians in the management of TMD, and these are sometimes considered as the first-line drugs of choice [46–48]. Besides, NSAID topical ointments are available as an alternative to lessen systemic absorption [40]. In a systematic review with meta-analysis conducted in 2017, Haggman et al. acknowledged the positive treatment effect of the muscle relaxant cyclobenzaprine for M-TMD, despite a lack of understanding of their long-term use and associated side effects [49]. Structurally similar to cyclobenzaprine are tricyclic antidepressants (TCAs). At a relatively low dose compared to treating depression, TCAs have been shown to be effective in reducing the frequency and intensity of pain arising from TMD; however, the numerous detrimental effects that ensue, notably dry mouth, fluid retention and cardiotoxicity, have limited its routine use [50]. Another potent class of central-acting drugs are opioids; however, their use has been discouraged to minimize central nervous system depression and physical dependence [51].

Although various medications are used in the management of TMD, there has been, generally, a paucity of evidence to support a standardized regimen and the best class of medications to be used [40,52]. Therefore, the clinical decision still lies mainly with the clinicians' experience and comfort [53], tailored to the patient's best individual needs.

4.1.2. Occlusal Splint

The use of occlusal splints has been reported to have improved the mandibular movement and pain for patients with TMD [39,43,54–56] and is considered a basic treatment for TMD [44,54,57,58]. Yet, there has been, generally, a scarcity of strong evidence to validate its efficacy in these aspects [42,59], especially in the long term [57,60,61]. Interestingly, Alkhytari et al. have conducted a systematic review on stabilization splints in 2018 and concluded that the patient-reported treatment satisfaction, including domains other than pain relief such as psychological well-being, was beyond that of a placebo effect. The authors also suggested that, apart from the pain scale, variables concerning patient-reported satisfaction should also be considered when evaluating the efficacy of different treatment modalities [62].

Different splint designs are available at present; the most commonly used ones are soft or hard stabilization splints, including Tanner appliance, Fox appliance, Michigan splint or centric relation appliance, anterior repositioning splint and anterior bite splint. When compared with TMD of joint origin, hard stabilization splints have been found to yield better outcomes in treating myogenous TMD [63]. They can aid in promoting the functional recovery of masticatory muscles [64] and restoring postural balance [65]. While mini-anterior splints (similar to anterior deprogramming splints in mechanism) have been suggested to be the most effective splint design to manage muscle pain in TMD, there is still a lack of high-level evidence [63]. On the other hand, its long-term use is not always advocated due to the possible adverse effects on occlusal stability from prolonged disocclusion of posterior teeth [66]. Moreover, some splints that incorporate biofeedback features, such as vibratory stimulus upon parafunctional occlusal load, have been reported to offer additional treatment benefits [67].

4.1.3. Physiotherapy

Physiotherapy, also termed physical therapy or exercise therapy, has been proposed to play a vital role, especially in the management of myofascial pain complaints for patients with TMD [8,48,68–72]. It can be subdivided into self-exercise conducted by patients at home or manual therapy by a trained practitioner [73]. Educations on various homecare strategies, for instance, massage of the masticatory muscles, jaw opening exercise and applications of moist heat pads [69,73] are most commonly offered as low-cost, useful modalities free of adverse consequences at an early phase. Evidence has also attached importance to postural correction of the head and neck in reducing pain and increasing jaw mobility [70,74], which might be related to the restoration of cervical lordosis [75]. Furthermore, some widely cited examples of manual therapies are post-isometric muscle relaxation and myofascial release. Although the aforementioned procedures have been more extensively documented in treating muscles of the trunk and limbs, these relaxation techniques have been reported to improve musculoskeletal functions even in the masticatory system, possibly by relief of muscle tension [76]. A diversity of approaches are available and have been shown to be beneficial in the treatment of TMD, including but not limited to mobilization, stretch, endurance exercise, etc., but as of today, none has been proven superior; more scientific evidence is called for to formulate a standardized protocol [73,77–79]. Additionally, there has been evidence suggesting that physiotherapy might be able to improve headache-associated symptoms from TMD [69].

4.1.4. Electrical Modalities

Apart from self-manipulation therapy, electrical modalities are becoming a more popular non-invasive treatment modality for relieving acute and chronic pain in TMD patients.

A. Transcutaneous Electrical Nerve Stimulation (TENS)

Transcutaneous electric nerve stimulation (TENS), has been used for millennia to relieve pain. It utilizes electrodes placed on the skin, which are connected to the unit via wires to achieve a targeted therapeutic goal [80] (Figure 1). Electrical impulses are generated to descend pain signals to the spinal cord and brain, stimulate the production of endorphins, relieve peripheral and neuropathic pain and relax muscles [81]. Not only pain and muscle tenderness can be relieved [82]; TENS also demonstrates benefits in improving masticatory function in TMD patients by improving their mouth opening and eventually increasing their biting force [83]. As seen from the photo, TENS is a small and portable device, often battery-operated, which can sometimes even fit into a pocket. Another additional benefit of TENS is that patients can self-apply the electrical pads themselves without assistance.

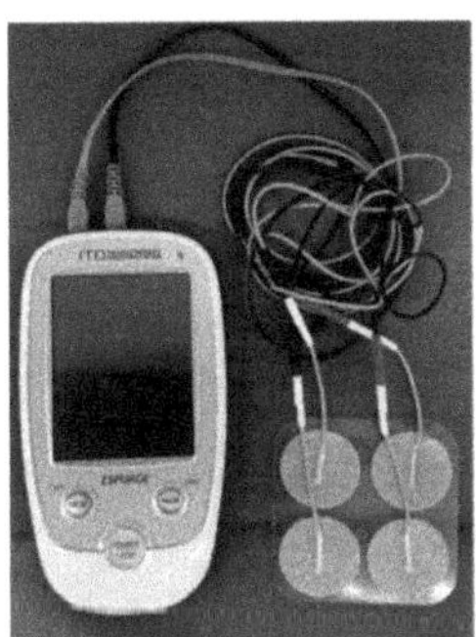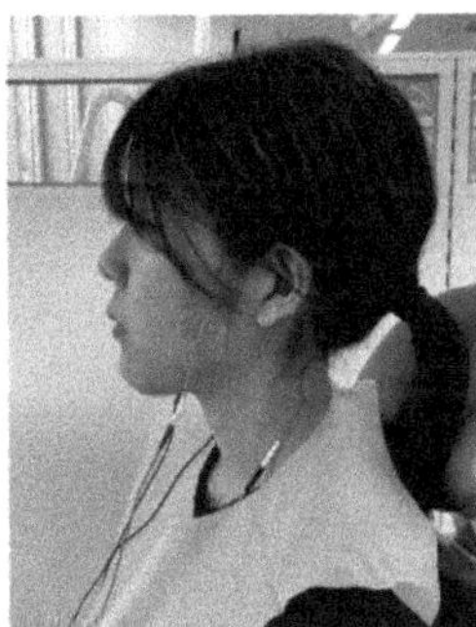

Figure 1. The transcutaneous electric nerve stimulation (TENS) device and its application on the masticatory muscles. (Figure from [83], https://pubmed.ncbi.nlm.nih.gov/33081336/#&gid=article-figures&pid=figure-1-uid-0 (accessed on 18 November 2022)).

B. Low-Level Laser Therapy (LLLT)

Among the various physical therapy modalities, low-level laser therapy (LLLT) has been placed under the spotlight because of its non-invasive, safe, easy application and short treatment time [84]. A recent systematic review has suggested that laser therapy has been particularly useful to treat muscle-related TMD pain among other rehabilitative approaches [85]. In addition, also termed photobiomodulation (PBM), LLLT involves a light source that emits no heat, sound or vibration but could affect the function of fibroblasts, facilitate repair and act as an anti-inflammatory agent [86]. One special feature of LLLT is that it does not make contact with skin and can be used even with wounds. In a systematic review and network meta-analysis conducted in 2022, Ren et al. explored the optimal wavelength range of laser application, affirming that laser therapy with a wavelength of 910–1100 nm was the most effective [87].

C. Therapeutic Ultrasound (US)

Similar to other electrical modalities, therapeutic US introduces energy to tissue cells, aiming to improve circulation to tissues and facilitate the healing process (Figure 2). There are two modes in therapeutic modes: continuous mode, which produces a deep heating effect (for chronic pain); and intermittent mode, which will not increase tissue temperature (for acute pain). US therapy can significantly reduce the pain and improve the functionality of the temporomandibular joint and mouth-opening limit for TMD patients in four weeks [88]. Yet, 2.63% of patients that had undergone ultrasound therapy had relapse and recurrence of pain [88]. Therefore, its long-term effectiveness is still inconclusive.

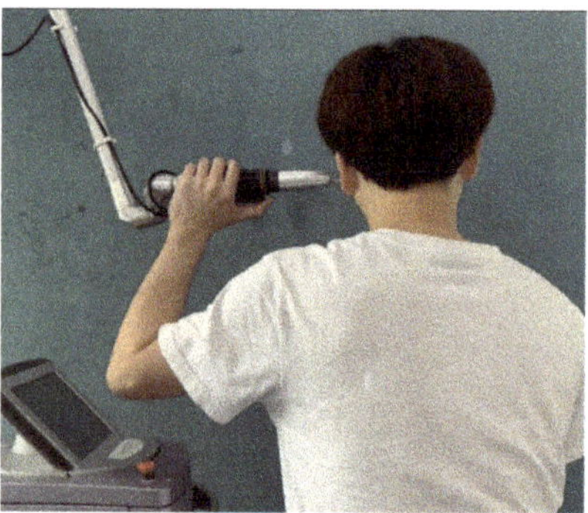

Figure 2. Therapeutic ultrasound (US). (Figure from [88], https://pubmed.ncbi.nlm.nih.gov/341408 03/#&gid=article-figures&pid=figure-1-uid-0 (accessed on 18 November 2022)).

While TENS, LLLT and therapeutic US are useful in pain reduction, systematic reviews and meta-analyses carried out in 2022 have shown that LLLT was the most effective in reducing pain among the three treatment modalities [87,89,90]; LLLT was found to be superior to TENS and was also proven to be better in reducing pain than therapeutic US [89]. Better results could be achieved with higher wavelengths, and wavelengths ranging from 910 nm to 1100 nm were recommended to treat TMD using LLLT [87].

D. Extracorporeal Shockwave Therapy (ESWT)

Radial and focused extracorporeal shockwave therapy (ESWT) is gaining popularity for treating musculoskeletal cases. It was hypothesized that the main biological effect on tissue treated by ESWT is an increase in the permeability of cell membranes and the release of several molecules stimulating tissue regeneration [91], such as vascular endothelial growth factor (VEGF), fibroblast growth factor (FGF) and the activation of the endothelial nitric oxide synthase (eNOS) with angiogenic effects [92]. Most importantly, it is believed that ESWT can modulate the release of anti-inflammatory mediators and endorphins that activate descending inhibitory system to relieve pain [93].

In 2022, Marotta et al. conducted an RCT to evaluate the efficacy of physical exercise, with or without radial extracorporeal shock wave therapy (rESWT), in patients with only muscular TMD. The findings of this pilot RCT suggested that rESWT combined with physical therapy could be effective in relieving pain and improving function in muscle-related TMD patients [94]. There was another interesting discovery from a qualitative analysis in 2022, which concluded that ESWT could facilitate both clinical and functional recovery in people with myofascial pain syndrome, but not for fibromyalgia [95]. Another point to note is that extracorporeal radial shockwave therapy combined with ultrasound-guided injection of lidocaine into MTrPs has been shown to be more effective for reducing pain and elastic stiffness in myofascial pain syndrome in the fourth week [96].

Despite considerable research that has supported the efficacy of ESWT on MTrPs in the trapezius muscle [97–100], little is known regarding its effect on mastication muscles. Considering this research gap, more clinical studies in this area are encouraged. Currently, our centre is conducting a randomized clinical trial on its use in patients presenting with myogenous pain (Figure 3); clinical improvements have been confirmed in our pilot study and its findings will be presented when the study is completed.

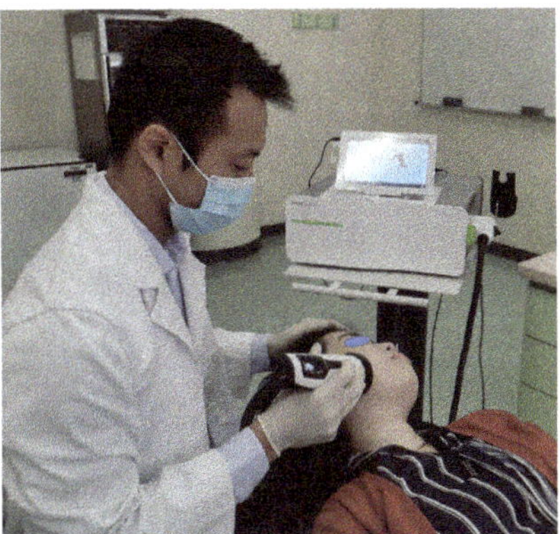

Figure 3. ESWT shows promising results for the treatment of MTrPs of masseter muscles in patients with myogenous TMD.

4.1.5. Psychological Intervention

Emerging evidence has shed light on the psychological component in both the aetiology and management of TMD [101]. There was an interesting finding by Nifosi et al. that patients suffering from M-TMD usually reported a higher stress level than those articular patients [102]. While it is logical to assume the causative relationship between anxiety and parafunctional behaviours, which could contribute to muscle hypertrophy [45], there are still insufficient data in the literature to elucidate the exact pain–psychopathology link. Nonetheless, in light of the multifactorial nature of TMD [45], psychosocial assessment has been integrated into its treatment to promote patients' mental well-being and potentially reduce harmful habits [103], leading to a rise in the popularity of a multimodal, biopsychosocial approach [63,103–105]. To deliver a comprehensive psychological-based therapy, a joint effort between TMD specialists and psychologists might be required; one notable, evidence-based example is cognitive behavioural therapy (CBT) [106–108].

However, more often than not, dental professionals might not be equipped with a solid background in psychological domains; therefore, counselling can be offered in the form of patient education, for example, on normal jaw function, suspected aetiological factors and reassurance of its benign nature, strategies with proven clinical efficacy [109–112].

4.2. Minimally Invasive Treatment

4.2.1. Dry Needling/Acupuncture

Dry needling (DN) or acupuncture are both treatment strategies targeted at the muscles, which have been widely used to treat a myriad of neuromusculoskeletal diseases, including myogenous TMD [113]. Although they differ slightly in their philosophy, western-based DN and acupuncture originating from traditional Chinese medicine both involve the insertion of long and fine needles into the MTrPs to relieve muscle tension and produce an analgesic effect [114–116]. Another distinguishing difference is that during DN, TrPs are repeatedly perforated internally and externally with the needle, such that a local twitch response might sometimes be observed [117]. Whilst insufficient data are available to compare the efficacy of these two needling techniques [118], they have been useful aids to manage pain and restore motions in some patients with muscle-originated TMD [48,114,119,120], despite the lack of clarity regarding the mechanism they are based upon [114]. However, due to the low quality of evidence and heterogenicity of the studies conducted, needling therapies have not been regarded as a first-line treatment for M-TMD [116,118].

4.2.2. Minimally Invasive Injections

A variety of drugs are available for injection therapy in the management of TMD [46]. They are usually classified by the mode of delivery: either as intra-articular injection, into the TMJ alone or as part of an arthrocentesis procedure [121,122]; or into the mastication muscles [123], and this article shall focus on the latter. Usually targeted at the MTrPs, intramuscular injections are also termed trigger point injections (TPIs). They can involve the use of local anaesthetic substances such as lidocaine, corticosteroids or botulinum toxin (BTX) [124]. In the past, lidocaine seems to be the preferred option for TPI due to its low cost [125]. Although limited information is available in the literature to compare the efficacies of different injectates, it has been observed that BTXs are increasingly popular in recent years [126], despite a lack of consensus on its clinical value [127–130]. For selected patients, especially those refractory to conventional conservative treatment measures, BTX injection might be a useful tool in modulating pain threshold and restoring motion [46,131–138]. It has been noted that the administration of BTX can also improve sleep bruxism [139], which has long been closely associated with the signs and symptoms of myogenous TMD. Note that the therapeutic effect of BTX injection is usually transient and repeated appointments are expected for long-term relief [140–142]. Concerns over side effects such as muscle paralysis and financial implications [143,144] have rendered it no more than an adjunct to other standard treatments [8].

5. Conclusions

Branched off from the umbrella term TMD, masticatory myalgia shares equally bewildering aetiology with multifaceted signs and symptoms. Aside from somatic cause, current evidence has recognized the role of psychosocial factors in its course of development. With this in mind, contemporary treatment approaches have placed more emphasis on bio-behavioural interventions, such as counselling therapy, alongside simultaneous conservative measures, to address various aspects of the issue in a multimodal fashion.

Note that this article is limited by the absence of meticulous meta-analysis in a systematic manner. It is our objective to provide an updated narrative overview of diagnosis and treatment modalities available for M-TMD. It has been widely accepted that the treatment philosophy of TMD remains empirical due to a paucity of knowledge in its pathophysiology. Further studies are needed to make sense of the clinical conundrum.

Author Contributions: Conceptualization, Y.Y.L.; writing—original draft preparation, N.H.Y.C. and C.K.I.; writing—review and editing, D.T.S.L. and Y.Y.L. All authors have read and agreed to the published version of the manuscript.

Funding: This research received no external funding.

Institutional Review Board Statement: Not applicable.

Informed Consent Statement: Not applicable.

Data Availability Statement: Not applicable.

Conflicts of Interest: The authors declare no conflict of interest.

References

1. Okeson, J.P. The classification of orofacial pains. *Oral Maxillofac. Surg. Clin. North Am.* **2008**, *20*, 133–144. [CrossRef] [PubMed]
2. Schiffman, E.; Ohrbach, R.; Truelove, E.; Look, J.; Anderson, G.; Goulet, J.P.; List, T.; Svensson, P.; Gonzalez, Y.; Lobbezoo, F.; et al. Diagnostic Criteria for Temporomandibular Disorders (DC/TMD) for Clinical and Research Applications: Recommendations of the International RDC/TMD Consortium Network* and Orofacial Pain Special Interest Group†. *J. Oral Facial Pain Headache* **2014**, *28*, 6–27. [CrossRef]
3. Valesan, L.F.; Da-Cas, C.D.; Réus, J.C.; Denardin, A.C.S.; Garanhani, R.R.; Bonotto, D.; Januzzi, E.; de Souza, B.D.M. Prevalence of temporomandibular joint disorders: A systematic review and meta-analysis. *Clin. Oral Investig.* **2021**, *25*, 441–453. [CrossRef]
4. Rongo, R.; Ekberg, E.; Nilsson, I.M.; Al-Khotani, A.; Alstergren, P.; Conti, P.C.R.; Durham, J.; Goulet, J.P.; Hirsch, C.; Kalaykova, S.I.; et al. Diagnostic criteria for temporomandibular disorders (DC/TMD) for children and adolescents: An international Delphi study-Part 1-Development of Axis, I. *J. Oral Rehabil.* **2021**, *48*, 836–845. [CrossRef]
5. Manfredini, D.; Guarda-Nardini, L.; Winocur, E.; Piccotti, F.; Ahlberg, J.; Lobbezoo, F. Research diagnostic criteria for temporomandibular disorders: A systematic review of axis I epidemiologic findings. *Oral Surg. Oral Med. Oral Pathol. Oral Radiol. Endod.* **2011**, *112*, 453–462. [CrossRef] [PubMed]

6. Dworkin, S.F.; Huggins, K.H.; LeResche, L.; Von Korff, M.; Howard, J.; Truelove, E.; Sommers, E. Epidemiology of signs and symptoms in temporomandibular disorders: Clinical signs in cases and controls. *J. Am. Dent. Assoc.* **1990**, *120*, 273–281. [CrossRef] [PubMed]

7. Kitsoulis, P.; Marini, A.; Iliou, K.; Galani, V.; Zimpis, A.; Kanavaros, P.; Paraskevas, G. Signs and symptoms of temporomandibular joint disorders related to the degree of mouth opening and hearing loss. *BMC Ear Nose Throat Disord.* **2011**, *11*, 5. [CrossRef] [PubMed]

8. Li, D.T.S.; Leung, Y.Y. Temporomandibular Disorders: Current Concepts and Controversies in Diagnosis and Management. *Diagnostics* **2021**, *11*, 459. [CrossRef]

9. Ferrillo, M.; Giudice, A.; Marotta, N.; Fortunato, F.; Di Venere, D.; Ammendolia, A.; Fiore, P.; de Sire, A. Pain Management and Rehabilitation for Central Sensitization in Temporomandibular Disorders: A Comprehensive Review. *Int. J. Mol. Sci.* **2022**, *23*, 12164. [CrossRef]

10. Sójka, A.; Stelcer, B.; Roy, M.; Mojs, E.; Pryliński, M. Is there a relationship between psychological factors and TMD? *Brain Behav.* **2019**, *9*, e01360. [CrossRef]

11. Kim, Y.K.; Kim, S.G.; Im, J.H.; Yun, P.Y. Clinical survey of the patients with temporomandibular joint disorders, using Research Diagnostic Criteria (Axis II) for TMD: Preliminary study. *J. Craniomaxillofac. Surg.* **2012**, *40*, 366–372. [CrossRef] [PubMed]

12. Yap, A.U.; Tan, K.B.; Chua, E.K.; Tan, H.H. Depression and somatization in patients with temporomandibular disorders. *J. Prosthet. Dent.* **2002**, *88*, 479–484. [CrossRef] [PubMed]

13. Toh, A.Q.J.; Chan, J.L.H.; Leung, Y.Y. Mandibular asymmetry as a possible etiopathologic factor in temporomandibular disorder: A prospective cohort of 134 patients. *Clin. Oral Investig.* **2021**, *25*, 4445–4450. [CrossRef] [PubMed]

14. Toh, A.Q.J.; Leung, Y.Y. The effect of orthognathic surgery on temporomandibular disorder. *J. Craniomaxillofac. Surg.* **2022**, *50*, 218–224. [CrossRef]

15. Leung, Y.Y.; Wang, R.; Wong, N.S.M.; Li, D.T.S.; Au, S.W.; Choi, W.S.; Su, Y.X. Surgical morbidities of sagittal split ramus osteotomy versus intraoral vertical ramus osteotomy for the correction of mandibular prognathism: A randomized clinical trial. *Int. J. Oral Maxillofac. Surg.* **2021**, *50*, 933–939. [CrossRef]

16. Peixoto, K.O.; Resende, C.; Almeida, E.O., Almeida-Leite, C.M.; Conti, P.C.R.; Barbosa, G.A.S.; Barbosa, J.S. Association of sleep quality and psychological aspects with reports of bruxism and TMD in Brazilian dentists during the COVID-19 pandemic. *J. Appl. Oral Sci.* **2021**, *29*, e20201089. [CrossRef]

17. Lavelle, E.D.; Lavelle, W.; Smith, H.S. Myofascial trigger points. *Anesth. Clin.* **2007**, *25*, 841–851. [CrossRef]

18. Dawson, A.; Ghafouri, B.; Gerdle, B.; List, T.; Svensson, P.; Ernberg, M. Effects of experimental tooth clenching on pain and intramuscular release of 5-HT and glutamate in patients with myofascial TMD. *Clin. J. Pain* **2015**, *31*, 740–749. [CrossRef]

19. Herpich, C.M.; Gomes, C.; Dibai-Filho, A.V.; Politti, F.; Souza, C.D.S.; Biasotto-Gonzalez, D.A. Correlation Between Severity of Temporomandibular Disorder, Pain Intensity, and Pressure Pain Threshold. *J. Manip. Physiol. Ther.* **2018**, *41*, 47–51. [CrossRef]

20. Fillingim, R.B.; Ohrbach, R.; Greenspan, J.D.; Knott, C.; Diatchenko, L.; Dubner, R.; Bair, E.; Baraian, C.; Mack, N.; Slade, G.D.; et al. Psychological factors associated with development of TMD: The OPPERA prospective cohort study. *J. Pain* **2013**, *14*, T75–T90. [CrossRef]

21. Jiménez-Silva, A.; Peña-Durán, C.; Tobar-Reyes, J.; Frugone-Zambra, R. Sleep and awake bruxism in adults and its relationship with temporomandibular disorders: A systematic review from 2003 to 2014. *Acta Odontol. Scand.* **2017**, *75*, 36–58. [CrossRef]

22. Alvarez, D.J.; Rockwell, P.G. Trigger points: Diagnosis and management. *Am. Fam. Physician* **2002**, *65*, 653–660.

23. van der Meer, H.A.; Bakke, M.; Schytz, H.W.; Madsen, B.K. Validation of the Temporomandibular Disorder Pain Screener in a Specialized Headache Center. *J. Oral Facial Pain Headache* **2021**, *35*, 150–156. [CrossRef]

24. Sahbaz, T.; Karacay, B.C. Investigation of temporomandibular disorders in patients with fibromyalgia syndrome: A case-control study. *J. Stomatol. Oral Maxillofac. Surg.* **2022**. [CrossRef] [PubMed]

25. Thambar, S.V.; Kulkarni, S.; Tesar, J.; Armstrong, S. Management of temporomandibular disorders in the primary care setting. *Med. J. Aust* **2021**, *214*, 412–414.e1. [CrossRef]

26. Ferneini, E.M. Temporomandibular Joint Disorders (TMD). *J. Oral Maxillofac. Surg.* **2021**, *79*, 2171–2172. [CrossRef] [PubMed]

27. Shah, J.P.; Thaker, N.; Heimur, J.; Aredo, J.V.; Sikdar, S.; Gerber, L. Myofascial Trigger Points Then and Now: A Historical and Scientific Perspective. *PM R* **2015**, *7*, 746–761. [CrossRef]

28. Lee, K.; Chon, S. Assessments of Muscle Thickness and Tonicity of the Masseter and Sternocleidomastoid Muscles and Maximum Mouth Opening in Patients with Temporomandibular Disorder. *Healthcare* **2021**, *9*, 1640. [CrossRef]

29. Al-Saleh, M.A.; Alsufyani, N.A.; Saltaji, H.; Jaremko, J.L.; Major, P.W. MRI and CBCT image registration of temporomandibular joint: A systematic review. *J. Otolaryngol. Head Neck Surg.* **2016**, *45*, 30. [CrossRef]

30. Derwich, M.; Mitus-Kenig, M.; Pawlowska, E. Interdisciplinary Approach to the Temporomandibular Joint Osteoarthritis-Review of the Literature. *Medicina* **2020**, *56*, 225. [CrossRef] [PubMed]

31. Shen, S.; Ye, M.; Wu, M.; Zhou, W.; Xu, S. MRI and DC/TMD Methods Analyze the Diagnostic Accuracy of the Change in Articular Disc of Temporomandibular Joint. *Comput. Math. Methods Med.* **2022**, *2022*, 1770810. [CrossRef]

32. Hu, S.; Shao, Z.; Deng, L. Clinical Manifestations, Imaging Features, and Pathogenic/Prognostic Risk Factors for Temporomandibular Disorders (TMD): A Case-Control Study Based on Psychogenic Factors of Patients. *Comput. Math. Methods Med.* **2022**, *2022*, 8279357. [CrossRef] [PubMed]

33. Diep, D.; Chen, K.J.Q.; Kumbhare, D. Ultrasound-guided interventional procedures for myofascial trigger points: A systematic review. *Reg. Anesth. Pain Med.* **2021**, *46*, 73–80. [CrossRef]

34. Lee, J.C.; Mitchell, A.W.; Healy, J.C. Imaging of muscle injury in the elite athlete. *Br. J. Radiol.* **2012**, *85*, 1173–1185. [CrossRef]

35. Hachisuka, R.; Taguchi, S.; Moriwaki, K.; Oshita, K.; Umeda, A.; Tsutsumi, Y.M. Subacute postoperative myofascial pain diagnosed and treated successfully by ultrasound: A case after laparoscopic hepatectomy. *JA Clin. Rep.* **2022**, *8*, 53. [CrossRef] [PubMed]

36. Sikdar, S.; Shah, J.P.; Gebreab, T.; Yen, R.H.; Gilliams, E.; Danoff, J.; Gerber, L.H. Novel applications of ultrasound technology to visualize and characterize myofascial trigger points and surrounding soft tissue. *Arch. Phys. Med. Rehabil.* **2009**, *90*, 1829–1838. [CrossRef] [PubMed]

37. Takla, M.K.; Razek, N.M.; Kattabei, O.; El-Lythy, M.A. A comparison between different modes of real-time sonoelastography in visualizing myofascial trigger points in low back muscles. *J. Man. Manip. Ther.* **2016**, *24*, 253–263. [CrossRef] [PubMed]

38. Perrot, S.; Cohen, M.; Barke, A.; Korwisi, B.; Rief, W.; Treede, R.D. The IASP classification of chronic pain for ICD-11: Chronic secondary musculoskeletal pain. *Pain* **2019**, *160*, 77–82. [CrossRef]

39. Zhang, S.H.; He, K.X.; Lin, C.J.; Liu, X.D.; Wu, L.; Chen, J.; Rausch-Fan, X. Efficacy of occlusal splints in the treatment of temporomandibular disorders: A systematic review of randomized controlled trials. *Acta Odontol. Scand.* **2020**, *78*, 580–589. [CrossRef]

40. Gil-Martínez, A.; Paris-Alemany, A.; López-de-Uralde-Villanueva, I.; La Touche, R. Management of pain in patients with temporomandibular disorder (TMD): Challenges and solutions. *J. Pain Res.* **2018**, *11*, 571–587. [CrossRef]

41. Kapos, F.P.; Exposto, F.G.; Oyarzo, J.F.; Durham, J. Temporomandibular disorders: A review of current concepts in aetiology, diagnosis and management. *Oral Surg.* **2020**, *13*, 321–334. [CrossRef] [PubMed]

42. Al-Moraissi, E.A.; Conti, P.C.R.; Alyahya, A.; Alkebsi, K.; Elsharkawy, A.; Christidis, N. The hierarchy of different treatments for myogenous temporomandibular disorders: A systematic review and network meta-analysis of randomized clinical trials. *Oral Maxillofac. Surg.* **2021**, *26*, 519–533. [CrossRef] [PubMed]

43. Wänman, A.; Marklund, S. Treatment outcome of supervised exercise, home exercise and bite splint therapy, respectively, in patients with symptomatic disc displacement with reduction: A randomised clinical trial. *J. Oral Rehabil.* **2020**, *47*, 143–149. [CrossRef] [PubMed]

44. Zhang, L.; Xu, L.; Wu, D.; Yu, C.; Fan, S.; Cai, B. Effectiveness of exercise therapy versus occlusal splint therapy for the treatment of painful temporomandibular disorders: A systematic review and meta-analysis. *Ann. Palliat. Med.* **2021**, *10*, 6122–6132. [CrossRef] [PubMed]

45. Melo, R.A.; de Resende, C.; Rêgo, C.R.F.; Bispo, A.S.L.; Barbosa, G.A.S.; de Almeida, E.O. Conservative therapies to treat pain and anxiety associated with temporomandibular disorders: A randomized clinical trial. *Int. Dent. J.* **2020**, *70*, 245–253. [CrossRef]

46. Andre, A.; Kang, J.; Dym, H. Pharmacologic Treatment for Temporomandibular and Temporomandibular Joint Disorders. *Oral Maxillofac. Surg. Clin. North Am.* **2022**, *34*, 49–59. [CrossRef]

47. Kurita Varoli, F.; Sucena Pita, M.; Sato, S.; Issa, J.P.M.; do Nascimento, C.; Pedrazzi, V. Analgesia Evaluation of 2 NSAID Drugs as Adjuvant in Management of Chronic Temporomandibular Disorders. *Sci. World J.* **2015**, *2015*, 359152. [CrossRef]

48. Jung, S.Y.; Kim, E.S.; Kim, K.W.; Cho, J.H.; Lee, Y.J.; Ha, I.H. Treatment for temporomandibular disorders in South Korea: A 9-year trend using cluster analysis. *J. Oral Rehabil.* **2022**, *49*, 691–700. [CrossRef]

49. Häggman-Henrikson, B.; Alstergren, P.; Davidson, T.; Högestätt, E.D.; Östlund, P.; Tranæus, S.; Vitols, S.; List, T. Pharmacological treatment of oro-facial pain–health technology assessment including a systematic review with network meta-analysis. *J. Oral Rehabil.* **2017**, *44*, 800–826. [CrossRef]

50. Huang-Lionnet, J.H.; Hameed, H.; Cohen, S.P. Chapter 53-Pharmacologic Management of Myofascial Pain. In *Essentials of Pain Medicine*, 4th ed.; Elsevier Inc.: Amsterdam, The Netherlands, 2018; pp. 475–484.e2.

51. Ouanounou, A.; Goldberg, M.; Haas, D.A. Pharmacotherapy in Temporomandibular Disorders: A Review. *J. Can. Dent. Assoc.* **2017**, *83*, h7.

52. Dalanon, J.; Ugalde, R.B.; Catibod, L.D.; Macaso, J.M.L.; Okura, K.; Matsuka, Y. Comparative analysis of education, awareness, and knowledge of dentists and physical therapists in the treatment of temporomandibular disorders. *Cranio* **2020**. *ahead-of-print*. [CrossRef]

53. Clarke, P.; Oluwajana, F.; James, M.; Crawford, C.; Foster-Thomas, E.; Allan, R. Temporomandibular disorders. Part 3: Pain and pharmacological therapy. *Dent. Update* **2022**, *49*, 453–460. [CrossRef]

54. Zhang, C.; Wu, J.Y.; Deng, D.L.; He, B.Y.; Tao, Y.; Niu, Y.M.; Deng, M.H. Efficacy of splint therapy for the management of temporomandibular disorders: A meta-analysis. *Oncotarget* **2016**, *7*, 84043–84053. [CrossRef] [PubMed]

55. Ramachandran, A.; Jose, R.; Tunkiwala, A.; Varma, R.B.M.; Shanmugham, A.; Nair, P.K.; Kumar, K.S.; Sam, L.M. Effect of deprogramming splint and occlusal equilibration on condylar position of TMD patients—A CBCT assessment. *Cranio* **2021**, *39*, 294–302. [CrossRef]

56. Seifeldin, S.A.; Elhayes, K.A. Soft versus hard occlusal splint therapy in the management of temporomandibular disorders (TMDs). *Saudi Dent. J.* **2015**, *27*, 208–214. [CrossRef] [PubMed]

57. Kuzmanovic Pficer, J.; Dodic, S.; Lazic, V.; Trajkovic, G.; Milic, N.; Milicic, B. Occlusal stabilization splint for patients with temporomandibular disorders: Meta-analysis of short and long term effects. *PLoS ONE* **2017**, *12*, e0171296. [CrossRef]

58. van Grootel, R.J.; Buchner, R.; Wismeijer, D.; van der Glas, H.W. Towards an optimal therapy strategy for myogenous TMD, physiotherapy compared with occlusal splint therapy in an RCT with therapy-and-patient-specific treatment durations. *BMC Musculoskelet Disord* **2017**, *18*, 76. [CrossRef] [PubMed]

59. Riley, P.; Glenny, A.M.; Worthington, H.V.; Jacobsen, E.; Robertson, C.; Durham, J.; Davies, S.; Petersen, H.; Boyers, D. Oral splints for temporomandibular disorder or bruxism: A systematic review. *Br. Dent. J.* **2020**, *228*, 191–197. [CrossRef]

60. Chen, H.M.; Liu, M.Q.; Yap, A.U.J.; Fu, K.Y. Physiological effects of anterior repositioning splint on temporomandibular joint disc displacement: A quantitative analysis. *J. Oral Rehabil.* **2017**, *44*, 664–672. [CrossRef]

61. Guo, Y.-n.; Cui, S.-j.; Zhou, Y.-h.; Wang, X.-d. An Overview of Anterior Repositioning Splint Therapy for Disc Displacement-related Temporomandibular Disorders. *Curr. Med. Sci.* **2021**, *41*, 626–634. [CrossRef]

62. Alkhutari, A.S.; Alyahya, A.; Rodrigues Conti, P.C.; Christidis, N.; Al-Moraissi, E.A. Is the therapeutic effect of occlusal stabilization appliances more than just placebo effect in the management of painful temporomandibular disorders? A network meta-analysis of randomized clinical trials. *J. Prosthet. Dent.* **2021**, *126*, 24–32. [CrossRef]

63. Al-Moraissi, E.A.; Farea, R.; Qasem, K.A.; Al-Wadeai, M.S.; Al-Sabahi, M.E.; Al-Iryani, G.M. Effectiveness of occlusal splint therapy in the management of temporomandibular disorders: Network meta-analysis of randomized controlled trials. *Int. J. Oral Maxillofac. Surg.* **2020**, *49*, 1042–1056. [CrossRef]

64. He, S.; Wang, S.; Song, F.; Wu, S.; Chen, J.; Chen, S. Effect of the use of stabilization splint on masticatory muscle activities in TMD patients with centric relation-maximum intercuspation discrepancy and absence of anterior/lateral guidance. *Cranio* **2021**, *39*, 424–432. [CrossRef]

65. Oliveira, S.S.I.; Pannuti, C.M.; Paranhos, K.S.; Tanganeli, J.P.C.; Laganá, D.C.; Sesma, N.; Duarte, M.; Frigerio, M.; Cho, S.C. Effect of occlusal splint and therapeutic exercises on postural balance of patients with signs and symptoms of temporomandibular disorder. *Clin. Exp. Dent. Res.* **2019**, *5*, 109–115. [CrossRef] [PubMed]

66. Zhao, M.; Wang, P.; Wang, H.; Li, X.; Bai, D.; Tian, Y. Diagnostic and treatment protocol for a patient with temporomandibular disorder using a stabilization splint and temporary anchorage devices. *Am. J. Orthod. Dentofac. Orthop.* **2021**, *159*, 666–681.e2. [CrossRef] [PubMed]

67. Bergmann, A.; Edelhoff, D.; Schubert, O.; Erdelt, K.J.; Pho Duc, J.M. Effect of treatment with a full-occlusion biofeedback splint on sleep bruxism and TMD pain: A randomized controlled clinical trial. *Clin. Oral Investig.* **2020**, *24*, 4005–4018. [CrossRef]

68. Incorvati, C.; Romeo, A.; Fabrizi, A.; Defila, L.; Vanti, C.; Gatto, M.R.A.; Marchetti, C.; Pillastrini, P. Effectiveness of physical therapy in addition to occlusal splint in myogenic temporomandibular disorders: Protocol of a randomised controlled trial. *BMJ Open* **2020**, *10*, e038438. [CrossRef] [PubMed]

69. van der Meer, H.A.; Calixtre, L.B.; Engelbert, R.H.H.; Visscher, C.M.; Nijhuis-van der Sanden, M.W.; Speksnijder, C.M. Effects of physical therapy for temporomandibular disorders on headache pain intensity: A systematic review. *Musculoskelet. Sci Pract.* **2020**, *50*, 102277. [CrossRef]

70. Dickerson, S.M.; Weaver, J.M.; Boyson, A.N.; Thacker, J.A.; Junak, A.A.; Ritzline, P.D.; Donaldson, M.B. The effectiveness of exercise therapy for temporomandibular dysfunction: A systematic review and meta-analysis. *Clin. Rehabil.* **2017**, *31*, 1039–1048. [CrossRef]

71. La Touche, R.; Martínez García, S.; Serrano García, B.; Proy Acosta, A.; Adraos Juárez, D.; Fernández Pérez, J.J.; Angulo-Díaz-Parreño, S.; Cuenca-Martínez, F.; Paris-Alemany, A.; Suso-Martí, L. Effect of Manual Therapy and Therapeutic Exercise Applied to the Cervical Region on Pain and Pressure Pain Sensitivity in Patients with Temporomandibular Disorders: A Systematic Review and Meta-analysis. *Pain Med.* **2020**, *21*, 2373–2384. [CrossRef]

72. Dinsdale, A.; Costin, B.; Dharamdasani, S.; Page, R.; Purs, N.; Treleaven, J. What conservative interventions improve bite function in those with temporomandibular disorders? A systematic review using self-reported and physical measures. *J. Oral Rehabil.* **2022**, *49*, 456–475. [CrossRef]

73. Shimada, A.; Ishigaki, S.; Matsuka, Y.; Komiyama, O.; Torisu, T.; Oono, Y.; Sato, H.; Naganawa, T.; Mine, A.; Yamazaki, Y.; et al. Effects of exercise therapy on painful temporomandibular disorders. *J. Oral Rehabil.* **2019**, *46*, 475–481. [CrossRef] [PubMed]

74. Armijo-Olivo, S.; Pitance, L.; Singh, V.; Neto, F.; Thie, N.; Michelotti, A. Effectiveness of Manual Therapy and Therapeutic Exercise for Temporomandibular Disorders: Systematic Review and Meta-Analysis. *Phys. Ther.* **2016**, *96*, 9–25. [CrossRef] [PubMed]

75. Derwich, M.; Gottesman, L.; Urbanska, K.; Pawlowska, E. Craniovertebral and Craniomandibular Changes in Patients with Temporomandibular Joint Disorders after Physiotherapy Combined with Occlusal Splint Therapy: A Prospective Case Control Study. *Medicina* **2022**, *58*, 684. [CrossRef] [PubMed]

76. Urbański, P.; Trybulec, B.; Pihut, M. The Application of Manual Techniques in Masticatory Muscles Relaxation as Adjunctive Therapy in the Treatment of Temporomandibular Joint Disorders. *Int. J. Env. Res. Public Health* **2021**, *18*, 12970. [CrossRef]

77. Morell, G.C. Manual therapy improved signs and symptoms of temporomandibular disorders. *Evid. Based Dent.* **2016**, *17*, 25–26. [CrossRef]

78. Espí-López, G.V.; Arnal-Gómez, A.; Cuerda del Pino, A.; Benavent-Corai, J.; Serra-Añó, P.; Inglés, M. Effect of Manual Therapy and Splint Therapy in People with Temporomandibular Disorders: A Preliminary Study. *J. Clin. Med.* **2020**, *9*, 2411. [CrossRef]

79. Eliassen, M.; Hjortsjö, C.; Olsen-Bergem, H.; Bjørnland, T. Self-exercise programmes and occlusal splints in the treatment of TMD-related myalgia-Evidence-based medicine? *J. Oral Rehabil.* **2019**, *46*, 1088–1094. [CrossRef]

80. Mokhtari, T.; Ren, Q.; Li, N.; Wang, F.; Bi, Y.; Hu, L. Transcutaneous Electrical Nerve Stimulation in Relieving Neuropathic Pain: Basic Mechanisms and Clinical Applications. *Curr. Pain Headache Rep.* **2020**, *24*, 14. [CrossRef]

81. Johnson, M.I.; Paley, C.A.; Jones, G.; Mulvey, M.R.; Wittkopf, P.G. Efficacy and safety of transcutaneous electrical nerve stimulation (TENS) for acute and chronic pain in adults: A systematic review and meta-analysis of 381 studies (the meta-TENS study). *BMJ Open* **2022**, *12*, e051073. [CrossRef]

82. Zhang, Y.; Liu, J.; Li, H.; Yan, Z.; Liu, X.; Cao, J.; Park, J.; Wilson, G.; Liu, B.; Kong, J. Transcutaneous auricular vagus nerve stimulation at 1 Hz modulates locus coeruleus activity and resting state functional connectivity in patients with migraine: An fMRI study. *Neuroimage Clin.* **2019**, *24*, 101971. [CrossRef]

83. Abe, S.; Miyagi, A.; Yoshinaga, K.; Matsuka, Y.; Matsumoto, F.; Uyama, E.; Suzuki, Y.; Oshima, M.; Okura, K.; Tanaka, E. Immediate Effect of Masticatory Muscle Activity with Transcutaneous Electrical Nerve Stimulation in Muscle Pain of Temporomandibular Disorders Patients. *J. Clin. Med.* **2020**, *9*, 3330. [CrossRef]

84. Xu, G.Z.; Jia, J.; Jin, L.; Li, J.H.; Wang, Z.Y.; Cao, D.Y. Low-Level Laser Therapy for Temporomandibular Disorders: A Systematic Review with Meta-Analysis. *Pain Res. Manag.* **2018**, *2018*, 4230583. [CrossRef]

85. Ferrillo, M.; Ammendolia, A.; Paduano, S.; Calafiore, D.; Marotta, N.; Migliario, M.; Fortunato, L.; Giudice, A.; Michelotti, A.; de Sire, A. Efficacy of rehabilitation on reducing pain in muscle-related temporomandibular disorders: A systematic review and meta-analysis of randomized controlled trials. *J. Back Musculoskelet. Rehabil.* **2022**, *35*, 916–921. [CrossRef]

86. Wang, L.; Liu, C.; Song, Y.; Wu, F. The effect of low-level laser irradiation on the proliferation, osteogenesis, inflammatory reaction, and oxidative stress of human periodontal ligament stem cells under inflammatory conditions. *Lasers Med. Sci.* **2022**. [CrossRef] [PubMed]

87. Ren, H.; Liu, J.; Liu, Y.; Yu, C.; Bao, G.; Kang, H. Comparative effectiveness of low-level laser therapy with different wavelengths and transcutaneous electric nerve stimulation in the treatment of pain caused by temporomandibular disorders: A systematic review and network meta-analysis. *J. Oral Rehabil.* **2022**, *49*, 138–149. [CrossRef] [PubMed]

88. Ba, S.; Zhou, P.; Yu, M. Ultrasound is Effective to Treat Temporomandibular Joint Disorder. *J. Pain Res.* **2021**, *14*, 1667–1673. [CrossRef]

89. Ansari, S.; Charantimath, S.; Lagali-Jirge, V.; Keluskar, V. Comparative efficacy of low-level laser therapy (LLLT) to TENS and therapeutic ultrasound in management of TMDs: A systematic review & meta-analysis. *Cranio* **2022**, *22*, 1–10. [CrossRef]

90. Chellappa, D.; Thirupathy, M. Comparative efficacy of low-Level laser and TENS in the symptomatic relief of temporomandibular joint disorders: A randomized clinical trial. *Indian J. Dent. Res.* **2020**, *31*, 42–47. [CrossRef] [PubMed]

91. Ryskalin, L.; Morucci, G.; Natale, G.; Soldani, P.; Gesi, M. Molecular Mechanisms Underlying the Pain-Relieving Effects of Extracorporeal Shock Wave Therapy: A Focus on Fascia Nociceptors. *Life* **2022**, *12*, 743. [CrossRef]

92. Aschermann, I.; Noor, S.; Venturelli, S.; Sinnberg, T.; Mnich, C.D.; Busch, C. Extracorporal Shock Waves Activate Migration, Proliferation and Inflammatory Pathways in Fibroblasts and Keratinocytes, and Improve Wound Healing in an Open-Label, Single-Arm Study in Patients with Therapy-Refractory Chronic Leg Ulcers. *Cell Physiol. Biochem.* **2017**, *41*, 890–906. [CrossRef]

93. Simplicio, C.L.; Purita, J.; Murrell, W.; Santos, G.S.; Dos Santos, R.G.; Lana, J. Extracorporeal shock wave therapy mechanisms in musculoskeletal regenerative medicine. *J. Clin. Orthop. Trauma* **2020**, *11*, S309–S318. [CrossRef]

94. Marotta, N.; Ferrillo, M.; Demeco, A.; Drago Ferrante, V.; Inzitari, M.T.; Pellegrino, R.; Pino, I.; Russo, I.; de Sire, A.; Ammendolia, A. Effects of Radial Extracorporeal Shock Wave Therapy in Reducing Pain in Patients with Temporomandibular Disorders: A Pilot Randomized Controlled Trial. *Appl. Sci.* **2022**, *12*, 3821. [CrossRef]

95. Paoletta, M.; Moretti, A.; Liguori, S.; Toro, G.; Gimigliano, F.; Iolascon, G. Efficacy and Effectiveness of Extracorporeal Shockwave Therapy in Patients with Myofascial Pain or Fibromyalgia: A Scoping Review. *Medicina* **2022**, *58*, 1014. [CrossRef] [PubMed]

96. Anwar, N.; Li, S.; Long, L.; Zhou, L.; Fan, M.; Zhou, Y.; Wang, S.; Yu, L. Combined effectiveness of extracorporeal radial shockwave therapy and ultrasound-guided trigger point injection of lidocaine in upper trapezius myofascial pain syndrome. *Am. J. Transl. Res.* **2022**, *14*, 182–196.

97. Kim, Y.H.; Bang, J.I.; Son, H.J.; Kim, Y.; Kim, J.H.; Bae, H.; Han, S.J.; Yoon, H.J.; Kim, B.S. Protective effects of extracorporeal shockwave on rat chondrocytes and temporomandibular joint osteoarthritis; preclinical evaluation with in vivo ^{99m}Tc-HDP SPECT and ex vivo micro-CT. *Osteoarthr. Cartil.* **2019**, *27*, 1692–1701. [CrossRef]

98. Schenk, I.; Vesper, M.; Nam, V.C. [Initial results using extracorporeal low energy shockwave therapy ESWT in muscle reflex-induced lock jaw]. *Mund Kiefer Gesichtschir.* **2002**, *6*, 351–355. [CrossRef] [PubMed]

99. Zhang, Q.; Fu, C.; Huang, L.; Xiong, F.; Peng, L.; Liang, Z.; Chen, L.; He, C.; Wei, Q. Efficacy of Extracorporeal Shockwave Therapy on Pain and Function in Myofascial Pain Syndrome of the Trapezius: A Systematic Review and Meta-Analysis. *Arch. Phys. Med. Rehabil.* **2020**, *101*, 1437–1446. [CrossRef] [PubMed]

100. Manafnezhad, J.; Salahzadeh, Z.; Salimi, M.; Ghaderi, F.; Ghojazadeh, M. The effects of shock wave and dry needling on active trigger points of upper trapezius muscle in patients with non-specific neck pain: A randomized clinical trial. *J. Back Musculoskelet. Rehabil.* **2019**, *32*, 811–818. [CrossRef] [PubMed]

101. Chung, J.; Lobbezoo, F.; van Selms, M.K.A.; Chattrattrai, T.; Aarab, G.; Mitrirattanakul, S. Physical, psychological and socio-demographic predictors related to patients' self-belief of their temporomandibular disorders' aetiology. *J. Oral Rehabil.* **2021**, *48*, 109–123. [CrossRef]

102. Nifosi, F.; Violato, E.; Pavan, C.; Sifari, L.; Novello, G.; Nardini, L.G.; Manfredini, D.; Semenzin, M.; Pavan, L.; Marini, M. Psychopathology and Clinical Features in an Italian Sample of Patients with Myofascial and Temporomandibular Joint Pain: Preliminary Data. *Int. J. Psychiatry Med.* **2007**, *37*, 283–300. [CrossRef]

103. de Freitas, R.F.; Ferreira, M.; Barbosa, G.A.; Calderon, P.S. Counselling and self-management therapies for temporomandibular disorders: A systematic review. *J. Oral Rehabil.* **2013**, *40*, 864–874. [CrossRef]

104. Costa, Y.M.; Porporatti, A.L.; Stuginski-Barbosa, J.; Bonjardim, L.R.; Conti, P.C.R. Additional effect of occlusal splints on the improvement of psychological aspects in temporomandibular disorder subjects: A randomized controlled trial. *Arch. Oral Biol.* **2015**, *60*, 738–744. [CrossRef]

105. Turp, J.C.; Jokstad, A.; Motschall, E.; Schindler, H.J.; Windecker-Getaz, I.; Ettlin, D.A. Is there a superiority of multimodal as opposed to simple therapy in patients with temporomandibular disorders? A qualitative systematic review of the literature. *Clin. Oral Implant. Res.* **2007**, *18*, 138–150. [CrossRef]

106. Matsuoka, H.; Chiba, I.; Sakano, Y.; Toyofuku, A.; Abiko, Y. Cognitive behavioral therapy for psychosomatic problems in dental settings. *BioPsychoSocial Med.* **2017**, *11*, 18. [CrossRef]

107. Dworkin, S.F.; Turner, J.A.; Mancl, L.; Wilson, L.; Massoth, D.; Huggins, K.H.; LeResche, L.; Truelove, E. A randomized clinical trial of a tailored comprehensive care treatment program for temporomandibular disorders. *J. Orofac. Pain* **2002**, *16*, 259–276.

108. Noma, N.; Watanabe, Y.; Shimada, A.; Usuda, S.; Iida, T.; Shimada, A.; Tanaka, Y.; Oono, Y.; Sasaki, K. Effects of cognitive behavioral therapy on orofacial pain conditions. *J. Oral Sci.* **2020**, *63*, 4–7. [CrossRef]

109. Donnarumma, V.; Michelotti, A.; Cimino, R.; Vollaro, S.; Cioffi, I. Short-term Effects of a First-Line Treatment Including Counseling and Self-Management Strategies on Chronic TMD Muscle Pain and Awake Bruxism in Women. *J. Oral Facial Pain Headache* **2022**, *36*, 36–48. [CrossRef]

110. Costa, Y.M.; Porporatti, A.L.; Stuginski-Barbosa, J.; Bonjardim, L.R.; Speciali, J.G.; Rodrigues Conti, P.C. Headache Attributed to Masticatory Myofascial Pain: Clinical Features and Management Outcomes. *J. Oral Facial Pain Headache* **2015**, *29*, 323–330. [CrossRef]

111. de Barros Pascoal, A.L.; de Freitas, R.F.C.P.; da Silva, L.F.G.; Oliveira, A.G.R.C.; Dos Santos Calderon, P. Effectiveness of Counseling on Chronic Pain Management in Patients with Temporomandibular Disorders. *J. Oral Facial Pain Headache* **2020**, *34*, 77–82. [CrossRef]

112. Magesty, R.A.; Silva, M.A.M.; Simões, C.A.S.C.; Falci, S.G.M.; Douglas-de-Oliveira, D.W.; Gonçalves, P.F.; Flecha, O.D. Oral health-related quality of life in patients with disc displacement with reduction after counselling treatment versus counselling associated with jaw exercises. *J. Oral Rehabil.* **2021**, *48*, 369–374. [CrossRef]

113. Pessoa, D.R.; Costa, D.R.; Prianti, B.M.; Costa, D.R.; Delpasso, C.A.; Arisawa EÂ, L.S.; Nicolau, R.A. Association of facial massage, dry needling, and laser therapy in Temporomandibular Disorder: Case report. *Codas* **2018**, *30*, e20170265. [CrossRef] [PubMed]

114. Fernández-Carnero, J.; La Touche, R.; Ortega-Santiago, R.; Galan-del-Rio, F.; Pesquera, J.; Ge, H.Y.; Fernández-de-Las-Peñas, C. Short-term effects of dry needling of active myofascial trigger points in the masseter muscle in patients with temporomandibular disorders. *J. Orofac. Pain* **2010**, *24*, 106–112.

115. Butts, R.; Dunning, J.; Pavkovich, R.; Mettille, J.; Mourad, F. Conservative management of temporomandibular dysfunction: A literature review with implications for clinical practice guidelines (Narrative review part 2). *J. Bodyw. Mov. Ther.* **2017**, *21*, 541–548. [CrossRef] [PubMed]

116. Machado, E.; Machado, P.; Wandscher, V.F.; Marchionatti, A.M.E.; Zanatta, F.B.; Kaizer, O.B. A systematic review of different substance injection and dry needling for treatment of temporomandibular myofascial pain. *Int. J. Oral Maxillofac. Surg.* **2018**, *47*, 1420–1432. [CrossRef] [PubMed]

117. Blasco-Bonora, P.M.; Martín-Pintado-Zugasti, A. Effects of myofascial trigger point dry needling in patients with sleep bruxism and temporomandibular disorders: A prospective case series. *Acupunct. Med.* **2017**, *35*, 69–74. [CrossRef]

118. Al-Moraissi, E.A.; Alradom, J.; Aladashi, O.; Goddard, G.; Christidis, N. Needling therapies in the management of myofascial pain of the masticatory muscles: A network meta-analysis of randomised clinical trials. *J. Oral Rehabil.* **2020**, *47*, 910–922. [CrossRef]

119. Dunning, J.; Butts, R.; Bliton, P.; Vathrakokoilis, K.; Smith, G.; Lineberger, C.; Eshleman, N.; Fernández-de-Las-Peñas, C.; Young, I.A. Dry needling and upper cervical spinal manipulation in patients with temporomandibular disorder: A multi-center randomized clinical trial. *Cranio* **2022**, 1–14. [CrossRef]

120. Simma, I.; Simma, L.; Fleckenstein, J. Muscular diagnostics and the feasibility of microsystem acupuncture as a potential adjunct in the treatment of painful temporomandibular disorders: Results of a retrospective cohort study. *Acupunct. Med.* **2018**, *36*, 415–421. [CrossRef]

121. Li, D.T.S.; Wong, N.S.M.; Li, S.K.Y.; McGrath, C.P.; Leung, Y.Y. Timing of arthrocentesis in the management of temporomandibular disorders: An integrative review and meta-analysis. *Int. J. Oral Maxillofac. Surg.* **2021**, *50*, 1078–1088. [CrossRef]

122. Leung, Y.Y.; Wu, F.H.W.; Chan, H.H. Ultrasonography-guided arthrocentesis versus conventional arthrocentesis in treating internal derangement of temporomandibular joint: A systematic review. *Clin. Oral Investig.* **2020**, *24*, 3771–3780. [CrossRef]

123. Wu, M.; Cai, J.; Yu, Y.; Hu, S.; Wang, Y.; Wu, M. Therapeutic Agents for the Treatment of Temporomandibular Joint Disorders: Progress and Perspective. *Front. Pharm.* **2020**, *11*, 596099. [CrossRef]

124. Bilici, I.Ş.; Emes, Y.; Aybar, B.; Yalçın, S. Evaluation of the effects of occlusal splint, trigger point injection and arthrocentesis in the treatment of internal derangement patients with myofascial pain disorders. *J. Cranio-Maxillo-Facial Surg.* **2018**, *46*, 916–922. [CrossRef]

125. Kamanli, A.; Kaya, A.; Ardicoglu, O.; Ozgocmen, S.; Zengin, F.O.; Bayik, Y. Comparison of lidocaine injection, botulinum toxin injection, and dry needling to trigger points in myofascial pain syndrome. *Rheumatol. Int.* **2005**, *25*, 604–611. [CrossRef]

126. Delcanho, R.; Val, M.; Nardini, L.G.; Manfredini, D. Botulinum Toxin for Treating Temporomandibular Disorders: What is the Evidence? *J. Oral Facial Pain Headache* **2022**, *36*, 6–20. [CrossRef]
127. Heir, G.M.D.M.D. The Efficacy of Pharmacologic Treatment of Temporomandibular Disorders. *Oral Maxillofac. Surg. Clin. North Am.* **2018**, *30*, 279–285. [CrossRef]
128. la Fleur, P.; Adams, A. CADTH Rapid Response Reports. In *Botulinum Toxin for Temporomandibular Disorders: A Review of Clinical Effectiveness, Cost-Effectiveness, and Guidelines*; Canadian Agency for Drugs and Technologies in Health: Ottawa, ON, Canada, 2020.
129. Muñoz Lora, V.R.M.; Del Bel Cury, A.A.; Jabbari, B.; Lacković, Z. Botulinum Toxin Type A in Dental Medicine. *J. Dent. Res.* **2019**, *98*, 1450–1457. [CrossRef]
130. Thambar, S.; Kulkarni, S.; Armstrong, S.; Nikolarakos, D. Botulinum toxin in the management of temporomandibular disorders: A systematic review. *Br. J. Oral Maxillofac. Surg.* **2020**, *58*, 508–519. [CrossRef]
131. Abboud, W.A.; Hassin-Baer, S.; Joachim, M.; Givol, N.; Yahalom, R. Localized myofascial pain responds better than referring myofascial pain to botulinum toxin injections. *Int. J. Oral Maxillofac. Surg.* **2017**, *46*, 1417–1423. [CrossRef]
132. De la Torre Canales, G.; Lorenzi Poluha, R.; Alvarez Pinzon, Y.N.; Rodrigues Conti, P.C.; Manfredini, D.; Sánchez-Ayala, A.; Rizzatti-Barbosa, C.M. Effects of Botulinum Toxin Type A on the Psychosocial Features of Myofascial Pain TMD Subjects: A Randomized Controlled Trial. *J. Oral Facial Pain Headache* **2021**, *35*, 288–296. [CrossRef]
133. De la Torre Canales, G.; Poluha, R.L.; Pinzón, N.A.; Da Silva, B.R.; Almeida, A.M.; Ernberg, M.; Manso, A.C.; Bonjardim, L.R.; Rizzatti-Barbosa, C.M. Efficacy of Botulinum Toxin Type-A I in the Improvement of Mandibular Motion and Muscle Sensibility in Myofascial Pain TMD Subjects: A Randomized Controlled Trial. *Toxins* **2022**, *14*, 441. [CrossRef]
134. Huamani, M.A.U.; Moreira, L.A.; AraÚJo, N.S.d.; Napimoga, M.H.; Junqueira, J.L.C.; Miranda, M.E. Use of botulinum toxin type a in temporomandibular disorder. *RGO Rev. Gaúcha De Odontol.* **2017**, *65*, 151–155. [CrossRef]
135. Meral, S.E.; Tüz, H.H.; Başlarlı, Ö. Evaluation of patient satisfaction after botulinum toxin A injection for the management of masticatory myofascial pain and dysfunction—A pilot study. *Cranio* **2021**, *39*, 12–16. [CrossRef] [PubMed]
136. Montes-Carmona, J.F.; Gonzalez-Perez, L.M.; Infante-Cossio, P. Treatment of Localized and Referred Masticatory Myofascial Pain with Botulinum Toxin Injection. *Toxins* **2020**, *13*, 6. [CrossRef] [PubMed]
137. Patel, A.A.; Lerner, M.Z.; Blitzer, A. IncobotulinumtoxinA Injection for Temporomandibular Joint Disorder. *Ann. Otol. Rhinol. Laryngol.* **2017**, *126*, 328–333. [CrossRef] [PubMed]
138. Sipahi Calis, A.; Colakoglu, Z.; Gunbay, S. The use of botulinum toxin-a in the treatment of muscular temporomandibular joint disorders. *J. Stomatol. Oral Maxillofac. Surg.* **2019**, *120*, 322–325. [CrossRef] [PubMed]
139. Beddis, H.; Pemberton, M.; Davies, S. Sleep bruxism: An overview for clinicians. *Br. Dent. J.* **2018**, *225*, 497–501. [CrossRef]
140. Bogucki, Z.A.; Kownacka, M. Clinical Aspects of the Use of Botulinum Toxin Type A in the Treatment of Dysfunction of the Masticatory System. *Adv. Clin. Exp. Med.* **2016**, *25*, 569–573. [CrossRef]
141. Christidis, N.; Omrani, S.; Fredriksson, L.; Gjelset, M.; Louca, S.; Hedenberg-Magnusson, B.; Ernberg, M. Repeated tender point injections of granisetron alleviate chronic myofascial pain—A randomized, controlled, double-blinded trial. *J. Headache Pain* **2015**, *16*, 104. [CrossRef]
142. Oksanen, E.; Männistö, V.; Kormi, E.; Vallioniemi, H.; Suojanen, J. Temporomandibular Disorder Patients Benefit From Intramuscular Botulinum Toxin Type a Injections. *J. Craniofac. Surg.* **2022**, *33*, 1159–1161. [CrossRef]
143. Chen, Y.W.; Chiu, Y.W.; Chen, C.Y.; Chuang, S.K. Botulinum toxin therapy for temporomandibular joint disorders: A systematic review of randomized controlled trials. *Int. J. Oral Maxillofac. Surg.* **2015**, *44*, 1018–1026. [CrossRef]
144. Patel, J.; Cardoso, J.A.; Mehta, S. A systematic review of botulinum toxin in the management of patients with temporomandibular disorders and bruxism. *Br. Dent. J.* **2019**, *226*, 667–672. [CrossRef]

 *diagnostics*

Article

Myogenous Temporomandibular Disorders: Diagnostic Concepts and Prospective Pilot Study on Extracorporeal Shockwave Therapy

Dion Tik Shun Li [1], Kar Yan Li [2] and Yiu Yan Leung [1,*]

1   Oral and Maxillofacial Surgery, Faculty of Dentistry, The University of Hong Kong, Hong Kong SAR, China
2   Clinical Research Centre, Faculty of Dentistry, The University of Hong Kong, Hong Kong SAR, China
*   Correspondence: mikeyyleung@hku.hk

Abstract: The aims of this article are to discuss the current, and potential future directions, in the diagnosis of myogenous temporomandibular disorders (M-TMD), as well as to report a pilot study to investigate the feasibility and clinical outcomes of extracorporeal shockwave therapy (ESWT) in the treatment of M-TMD. Forty-one adult patients presented with M-TMD were recruited into the study and randomized into two groups: Group 1 received ESWT treatment, whereas Group 2 received placebo treatment. The variables investigated were pain, measured by a numerical rating scale (NRS) and mouth opening. Twenty-six patients (Group 1: n = 14, mean age = 45.3 (16.7) years; Group 2: n = 12, mean age = 46.8 (19.7) years) completed 1-year follow up and were included into the final analysis. In both groups, reduction in pain and increase in MO (unassisted maximum, assisted maximum, and pain-free) were seen at post-treatment 1 year. There were more reduction in pain and increase in all MO in Group 1 than Group 2, but statistical significance was not detected. No major complications were encountered in this study. Although significant differences were not seen between groups, this prospective pilot study provided preliminary evidence that ESWT is safe and potentially beneficial in the treatment of M-TMD.

Keywords: temporomandibular disorders; extracorporeal shockwave therapy; maximal mouth opening; parallel group design

 check for updates

Citation: Li, D.T.S.; Li, K.Y.; Leung, Y.Y. Myogenous Temporomandibular Disorders: Diagnostic Concepts and Prospective Pilot Study on Extracorporeal Shockwave Therapy. *Diagnostics* **2023**, *13*, 51. https://doi.org/10.3390/diagnostics13010051

Academic Editor: Luis Eduardo Almeida

Received: 7 December 2022
Revised: 20 December 2022
Accepted: 22 December 2022
Published: 24 December 2022

1. Introduction

Temporomandibular disorders (TMD) are a group of conditions related to the impaired function of the temporomandibular joints (TMJ) and the associated neuro-muscular system [1]. Common complaints of TMD include clicking in the joint, pain in the TMJs or masticatory muscle and limited mouth opening, which could affect daily functions such as speech and mastication. The origin of the pain and dysfunction could be from the joint, muscles of mastication, or a combination of the two [2].

Since TMD of arthrogenous and myogenous nature may have different etiologies, it is important to differentiate between the two in the clinical setting, as the management approach of these conditions may be different [3,4]. Myogenous TMD (M-TMD) is known to mainly affect adult women from age 25–45 years old [5], and may have a higher tendency to seek treatment than those with TMD of arthrogenous origin [6]. Symptoms of M-TMD may include a moderate dull, pressing pain in the masticatory muscle which may become a sharper and more intense pain upon provocation and function [7]. It is known that apart from somatic causes, psychosocial factors may be responsible for the course of development of M-TMD among others [8].

Various treatment options for M-TMD are available, such as jaw exercises [9], medications [10], splint therapy [11], dry-needling [12], botox injection [13], cognitive behavioural therapy [14], and self-care instructions. While clinicians may find favourable outcomes in some of these options, there are those patients that are not fit for a particular treatment.

For example, good patient compliance is required for jaw exercises and occlusal splint therapy, and that occlusal splint therapy may not be appropriate in those who also have obstructive sleep apnea (OSA) requiring continuous positive airway pressure (CPAP). In addition, some of the treatment options might produce undesirable effects, such as possible change in facial shape in the case of botox injection.

Shockwave is a propagating disturbance of great amplitude which travels in a medium and is faster than the speed of sound. First described in the 1980s for the treatment of urolithiasis [15], extracorporeal shockwave therapy (ESWT) has since shown promising results in conditions such as plantar fasciitis [16,17], erectile dysfunction [18], spasticity in post-stroke patients [19], Achilles tendinopathy [19], and chronic calcific tendinitis of the shoulder [20]. Although the exact mechanism of ESWT in its therapeutic applications is unknown, it is speculated that it has an effect on wound healing [21,22]. Recently, ESWT for the treatment of TMD has gained interest. Initial results revealed beneficial outcomes in the treatment of muscle reflex-induced lock jaw with ESWT [23]. ESWT was shown to produce a protective effect on cartilage and subchondral bone structures in the rat model with temporomandibular joint osteoarthritis [24]. In a recent prospective study comparing the effects of ESWT with ultra-short wave (UW) in the treatment of TMD, ESWT resulted in improved response in terms of pain reduction and increased mouth opening in the short-term [25]. However, there are no studies that compare the efficacy and safety of ESWT compared with conventional treatments for M-TMD, and whether it can produce a long-term benefit is unknown at this time.

The aims of this article are to discuss the diagnostic concepts in M-TMD, and to report a prospective pilot study is to assess the feasibility and safety of ESWT in the treatment of M-TMD, and to compare the efficacy of ESWT and placebo therapy in terms of changes in pain score and mandibular function.

Diagnostic Concepts

Currently, the most accepted diagnostic tool for TMD is the Diagnostic Criteria for Temporomandibular Disorders (DC/TMD), which is used for both research and clinical purposes [1]. For M-TMD, diagnosis may include myalgia, local myalgia, myofascial pain, and myofascial pain with referral, which can involve the temporalis, masseter, and other muscles [1] (Table 1). At present, the standard for diagnosis of M-TMD mainly involves clinical examination and history taking, such as palpation of affected muscles and measurement of mandibular function, as opposed to TMD of arthrogenous origin in which imaging may also play a significant role in diagnosis. Magnetic resonance imaging (MRI) is considered the gold standard in the diagnosis of arthrogenous TMD, since disc abnormalities in location and morphology and presence of joint effusion could be readily assessed [26–28]. In addition, cone-beam computed tomography (CBCT) may be used to assess any bony pathologies of the mandibular condyle as well as the glenoid fossa [29].

Table 1. Classification of myogenous temporomandibular disorders (M-TMD) according to the Diagnostic Criteria for Temporomandibular Disorders (DC/TMD).

Classification of M-TMD	Clinical Findings
Myalgia	Familiar pain in the masseter and temporalis upon palpation or mouth opening
Local myalgia	Familiar pain in the masseter and temporalis localized to the site of palpation
Myofascial pain	Pain in the masseter and temporalis spreading beyond the site of palpation but within the confines of the muscle being palpated
Myofascial pain with referral	Pain in the masseter and temporalis beyond the confines of the muscle being palpated

At present, less is known about the imaging approach to the diagnosis of M-TMD. It has been suggested that M-TMD could be caused by injury to the masticatory muscles due to repeated strain from parafunctional habits, resulting in myofascial trigger points [30,31]. A myofascial trigger point has been described as a hyperirritable spot within a taut band of skeletal muscle, which may be painful to palpation and may also result in referred pain [31].

Understanding and locating such myofascial trigger points may be clinically important, as it has been shown that various invasive treatments, such as dry-needling and injection of platelet-rich plasma, may be useful to alleviate symptoms arising from such myofascial trigger points in the masseter muscle [32]. However, the use of imaging modalities is not routinely carried out in the management of M-TMD due to there being insufficient literature to support its application.

Although taut band, which may house myofascial trigger points, is readily palpable by a trained clinician, the detection of those on imaging is often less than straightforward [33]. Although MRI has been suggested to be useful in locating such taut bands in various muscles such as the trapezius [33,34], it is expensive, inconvenient, invasive in the case where contrast agent such as gadolinium is used, and has not been shown to be useful in the muscles of mastication. Another diagnostic imaging modality which has been proposed is ultrasonography (US) which may be more cost-effective, convenient and accessible. Various reports have described the efficacy of US in the identification of myofascial trigger points in muscles such as the lower back and trapezius muscle [35–38]. However, to the best knowledge of the authors, there are no reports on the identification of taut bands and myofascial trigger points in M-TMD using imaging modalities such as US. A clinical trial on the detection of myofascial trigger points using US in the management of M-TMD is currently underway at the authors centre in an attempt to fill such knowledge gap. The following sections of this paper will focus on a prospective pilot study on ESWT in the management of M-TMD.

2. Materials and Methods

This prospective pilot study was designed according to the CONSORT 2010 statement. Ethical approval was obtained from the Institutional Review Board of the University of Hong Kong/Hospital Authority Hong Kong West Cluster (HKU/HA HKW IRB) (IRB Reference Number: UW 20-704) prior to the start of this study. Written informed consent was obtained from participating subjects.

2.1. Study Design

This was a parallel-grouped clinical trial with balanced randomization (1:1).

2.2. Participants

Ethnic Chinese adults presented to the Discipline of Oral and Maxillofacial Surgery, Faculty of Dentistry, the University of Hong Kong for myogenous temporomandibular disorders were considered for recruitment into the current study:

2.2.1. Inclusion Criteria

- At least 16 years of age;
- Pain in the masticatory muscles, headache attributed to TMD, with or without limited mouth opening and pain in the TMJ.

2.2.2. Exclusion Criteria

- Pain in the TMJ only and not involving muscles of mastication;
- Active infection in the TMJ region;
- Systemic rheumatic diseases;
- Significant systemic diseases, such as uncontrolled hypertension, history of stroke within 6 months, and unstable angina;
- Craniofacial syndromes;
- Previous operations in the TMJ.

After an initial clinical examination and confirmation of the diagnosis of M-TMD (DTSL), the patients were prescribed a 2-week course of non-steroidal anti-inflammatory drugs (NSAIDs) (Ibuprofen 400 mg TDS), or paracetamol (500 mg QID) if NSAIDs were contraindicated. If the clinical symptoms were refractory to medication (no reduction in pain score), the patients were then be recruited into the study.

2.3. Pre-Treatment Assessment
2.3.1. Clinical Diagnosis

The clinical diagnosis was based on the Diagnostic Criteria for Temporomandibular Disorders (DC/TMD) [1]. Any type of pain in the masticatory muscle (myalgia, tendonitis, myositis, spasm) and any headache attributed to TMD were recorded.

2.3.2. Assessment of Pain

Pre-treatment pain symptoms, both at rest and during mandibular movement, was measured with a 11-point (0–10) numerical rating scale (NRS) adopted from the Graded Chronic Pain Scale [39], with 0 indicating no pain, while 10 indicates maximum pain.

2.3.3. Assessment of TMJ Function

Mouth opening (MO, pain-free, unassisted maximum, and assisted maximum, measured with a ruler between the incisal edges of the upper and lower incisors, minus the overbite), was measured in millimetres in the same way using the midpoints of the upper and lower incisors as references.

2.4. Interventions

The two arms of intervention are ESWT versus placebo. After palpation and identification of the region of pain, the patients were blinded with regards to which group they had been allocated to, and were treated in the following manner:

2.4.1. Group 1: Extracorporeal Shockwave Therapy

In the ESWT group, Focused ESWT (DUOLITH® SD1 T-TOP, Storz Medical) was applied at 0.15 mJ/mm^2 and stand-off II as per manufacturer's recommendations for craniomandibular dysfunction (CMD) to the painful side of the masseter muscle by a single operator (DTSL), for three sessions delivered at one-week intervals. At each session, 500 pulses were delivered to the masseter muscle.

2.4.2. Group 2: Placebo

In the placebo group, the handpiece of the ESWT was connected to a placebo stand-off with zero energy output so that no shockwave was transmitted to the patient. The procedure was carried out in the same way as in the ESWT. The patients were treated once a week for 3 weeks.

2.5. Outcomes

The primary outcome measure was pain symptoms in 6 weeks, as measured with a numerical rating scale (NRS). The secondary outcome measure was TMJ function.

2.5.1. Assessment of Pain

Post-treatment assessment of pain symptoms was measured with an NRS 1 week after each treatment session and at 6 weeks, 3 months, 6 months and 1 year after the first treatment session.

2.5.2. Assessment of Mouth Opening

Post-treatment assessment of pain-free mouth opening was carried out 1 week after each treatment session and at 6 weeks, 3 months, 6 months and 1 year after the first treatment session. Post-treatment assessment of maximum unassisted and maximum assisted mouth opening was performed at post-treatment 1 year.

2.5.3. Complications

All intra-operative, immediate, early post-operative, and late post-operative complications were recorded.

2.6. Randomization

Recruited patients were randomized into one of the two study groups by a simple randomization procedure. Using a computer program, a randomization table was generated. The allocation sequence was kept concealed in sequentially numbered, opaque and sealed envelopes. Upon obtaining the study consent from the participants, the surgeon in-charge would open the sealed envelope containing the allocation sequence.

2.7. Statistical Methods

Statistical analysis was performed using SPSS 28 software (IBM Corp., New York, NY, USA). For continuous variables, Shapiro-Wilk normality test was performed to test if the data followed normal distribution. For testing the differences at the same time interval between groups, independent t-test (or Mann-Whitney test if normality of the data was not fulfilled) was used. Comparison between baseline and other follow-up time points were performed with the paired-sample t-test (or Wilcoxon Signed Ranks test if normality of the data was not fulfilled). For the analysis of multiple comparisons at other follow-up time points and multiple comparisons between baseline and other follow-up time points, Bonferroni correction was used. $p < 0.05$ was considered statistically significant.

3. Results

Patients whose symptoms of M-TMD were not alleviated by NSAIDs were recruited into the study. A rolling recruitment and randomization strategy were employed in anticipation of a sizable dropout due to the COVID-19 pandemic. A total of 41 patients were initially recruited into the study. Of these, 21 patients were allocated to Group 1 (ESWT) and 20 patients were allocated to Group 2 (placebo). Loss of follow-up occurred in various time-points. The final number of patients who completed the 1-year follow-up schedule included 26 patients: 14 patients in Group 1 and 12 patients in Group 2 (Figure 1).

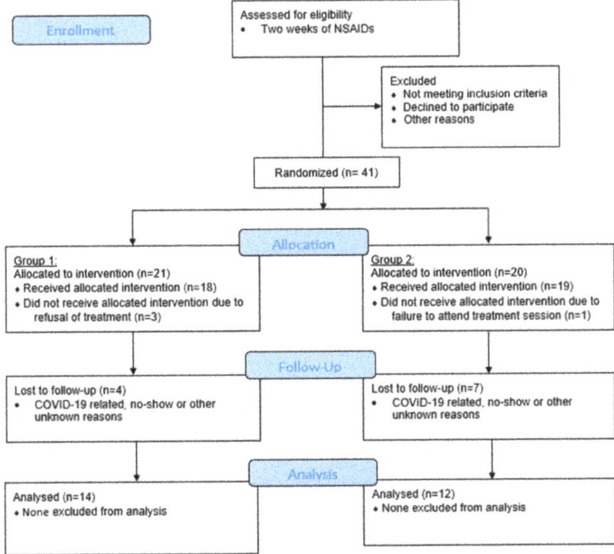

Figure 1. CONSORT flow diagram (version 2010) of subject enrolment, allocation, and follow-up. Group 1: extracorporeal shockwave therapy (ESWT); Group 2: placebo.

3.1. Patient Characteristics

Demographic characteristics of the final study sample is shown in Table 2. Male patients consisted of 14.3% in Group 1 (n = 2), and 25% in Group 2 (n = 3). The mean age at the time of recruitment was 45.3 (16.7) and 46.8 (19.7) years, respectively, for Group 1 and 2. The mean duration of pain in months at recruitment was 33.5 (36.2) and 42.6 (43.1) for Group 1 and 2, respectively. Other variables are presented in Table 2. There was no significance in any of the demographic characteristics between the two groups.

Table 2. Demographic characteristics of the final analysed sample.

	Group 1: ESWT	Group 2: Placebo
Sample size, n	14	12
Male, n (%)	2 (14.3)	3 (25)
Female, n (%)	12 (85.7)	9 (75)
Mean age in years (SD)	45.29 (16.7)	46.75 (19.7)
Duration of pain in months (SD)	33.5 (36.2)	42.58 (43.1)
Pain (NRS) (SD)	7.07 (1.7)	5.67 (1.5)
MO (pain free, mm) (SD)	31.57 (11.9)	30.08 (7.9)
MO (max unassisted, mm) (SD)	37.5 (8.1)	37.83 (7.4)
MO (max assisted, mm) (SD)	41 (7.9)	39.58 (7.5)
Painful conditions (%)		
Arthralgia	5 (35.7)	7 (58.3)
Myalgia	10 (71.4)	10 (83.3)
Myofascial pain with referral	4 (28.6)	2 (16.7)
Headache attributed to TMD	6 (42.9)	3 (25)
Non-painful conditions (%)		
DDWR	6 (42.9)	5 (41.7)
DDWR with intermittent locking	1 (7.1)	3 (25)
DDWOR with limited mouth opening	0	1 (8.3)

ESWT, extracorporeal shockwave therapy; SD, standard deviation; NRS, numerical rating scale; MO, mouth opening; mm, millimetres; TMD, temporomandibular disorders; DDWR, disc displacement with reduction; DDWOR, disc displacement without reduction.

3.2. Pain

Figure 2 shows the progression of clinical outcomes (pain, unassisted maximal mouth opening, assisted maximal mouth opening, and pain-free mouth opening), based on the raw data. A normality test showed that pain did not follow a normal distribution ($p < 0.05$) and thus non-parametric tests were performed.

At baseline (T0), there was no significant difference in pain between Group 1 and Group 2 ($p = 0.063$). There was a greater reduction in pain in the ESWT group than the placebo group at subsequent time points. However, there was no significant difference between the two groups in terms of pain at any follow up time points after Bonferroni correction. Within each group, pain dropped over time significantly compared to baseline. Within Group 1, after Bonferroni correction, significant differences were observed between T0 and T1 ($p = 0.024$), between T0 and T2 ($p = 0.013$), between T0 and T3 ($p = 0.020$), between T0 and T4 ($p = 0.002$), between T0 and T5 ($p = 0.020$) and between T0 and T6 ($p = 0.013$). However, within Group 2, after Bonferroni correction, significant differences were only detected between T0 and T3 ($p = 0.042$), between T0 and T4 ($p = 0.019$), and between T0 and T5 ($p = 0.0030$) (Figure 2).

3.3. Mouth Opening

A normality test showed that unassisted MO did not follow a normal distribution ($p < 0.05$) and thus non-parametric tests were performed. At baseline (T0), there were no significant differences in unassisted maximum MO ($p = 0.816$), assisted maximum MO ($p = 0.659$) and pain-free MO ($p = 0.725$) between the two groups. Then, at all follow-up time points, after Bonferroni correction, no significant differences were detected in unassisted maximum MO, assisted maximum MO, and pain-free MO between the two groups. Within each group for unassisted maximum MO, the only significant difference was found between T0 and T2 in Group 1 ($p = 0.029$) and between T0 and T4 in group 2 ($p = 0.039$) after Bonferroni correction. No significance time changes in assisted maximal MO and pain-free were found within both groups (Figure 2).

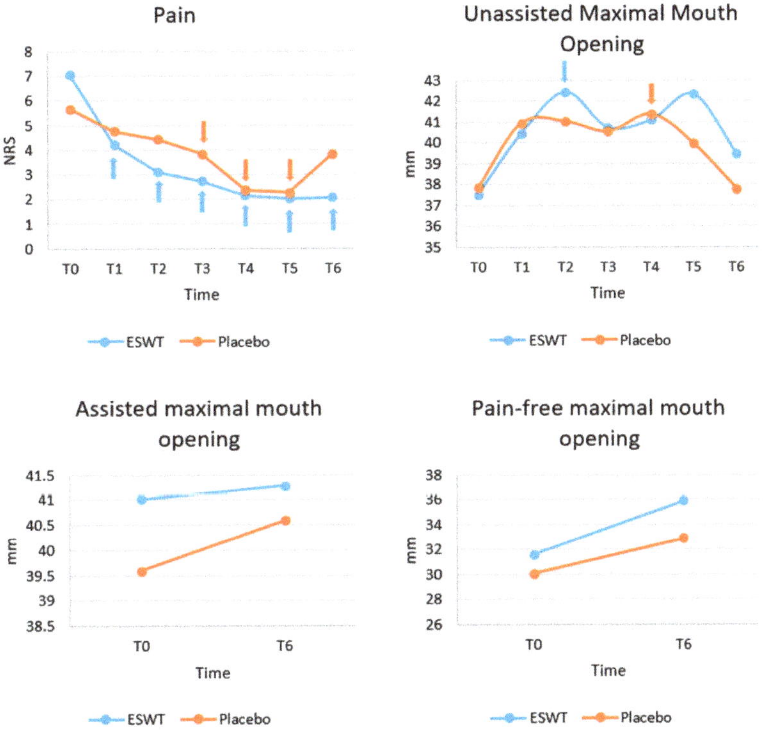

Figure 2. Progression of clinical variables (pain, unassisted maximum mouth opening, assisted maximum mouth opening, pain-free mouth opening) at different time points. Although greater reduction in pain and increase in mouth opening was seen in Group 1, there was no statistically significant difference of any variables at any time point between the two groups. For within group analyses, those values with significant difference compared to baseline (T0) are marked with arrows with corresponding colours. NRS, numerical rating scale; MO, mouth opening; T0, baseline; T1, 1 week after the first session; T2, 2 weeks after the first session; T3, 3 weeks after the first session; T4, 6 weeks after the first session; T5, post-treatment 3 months; T6, post-treatment 1 year.

3.4. Complications

One patient in Group 1 (male, age 40 years) complained of increased pain which radiated to the temporalis and neck one day after the application of ESWT. He was given paracetamol and the pain subsided the next day. There were no other complications observed.

4. Discussion

The diagnosis and management of M-TMD are not always straightforward due to the little understanding of the value of other investigations apart from clinical examination. At present, there is an obvious gap in the literature concerning the use of imaging in the diagnosis of M-TMD. Moreover, there is no consensus on which treatment options are superior to others, as no single treatment modality has consistently provided predictable outcomes [40]. Thus, the management for any individual patient may often be based on a trial-and-error approach and may comprise of a combination of treatment options. Apart from the potential clinical efficacy that a particular treatment modality may be able to deliver, other important considerations in selecting treatment options include safety, convenience, cost-effectiveness, patient compliance, and the possibility to be combined with other treatment modalities.

The result of this prospective pilot study has shown that ESWT is safe and is a potentially beneficial treatment option in the management of myogenous TMD. Reduction in pain and improvement of mouth opening were found to be superior in the ESWT group compared to the placebo group, though these findings were not statistically significant and was likely attributed to the small sample size in this pilot study.

Currently, there are various treatment options for M-TMD that are commonly employed by clinicians and may be considered conventional options. These include occlusal splints [41,42], physical therapy [11], counselling therapy [43,44], and botulinum toxin injection [12,13,45]. Other treatments that have been mentioned in the literature but may be less common include dry-needling [12,40], low level laser [46], and acupuncture [12,47]. For pain relief of myofascial trigger points in the trapezius muscle, dry-needling and low-level laser have shown promising results [48]; however, whether this can be applied to myofascial trigger points in M-TMD is unknown. In a recent systematic review and network meta-analysis of randomized clinical trials, it was found that manual therapy, counselling therapy, occlusal splint, and botulinum toxin injection may have a positive effect in the management of M-TMD in both the short and intermediate term [40]. On the other hand, another systematic review has found that placebo effect may be responsible for the positive outcomes seen in some other treatment options, such as acupuncture and dry needling [49]. However, the evidence of these findings is mostly of low quality due to the limitations of the studies included.

ESWT may be an emerging novel treatment modality in the management of M-TMD and may be offered in conjunction of other available treatments. ESWT may have a positive effect on wound healing [21,22] which may be beneficial as an additional treatment option for M-TMD. Although ESWT has been describe as a therapeutic option for multiple applications [16,16–20], to the authors' best knowledge, there are no other studies in the literature that compare ESWT with other conventional treatment options of M-TMD, nor any other studies that describe its safety and feasibility. A comparative study with a 4-week follow up has shown that ESWT may be beneficial in the management of temporomandibular joint disorder, when compared to ultrashort wave (UW) applied to the TMJ, in terms of reduction in pain and improvement of mandibular function [25]. The results of this study suggest that positive outcomes may also be applicable in the treatment of M-TMD and when compared to conventional treatments of M-TMD. However, future prospective studies with a larger sample size may be able to detect any statistical significance.

The current prospective pilot study provides preliminary information regarding ESWT in the treatment of M-TMD; however, there are several limitations. First of all, the sizable dropout rate may represent potential bias in this study, as patients with different traits or resultant clinical outcomes may be more prone to dropout. Moreover, as M-TMD may represent a chronic pain syndrome with patients consulting multiple clinicians and have attempted various treatment options, many patients recruited in this study had received prior treatments for M-TMD. In addition, as a pilot study, the post-operative accessor (DTSL) was the same as the treating clinician and thus was not blinded. Another notable finding was that significant clinical improvement was also seen in the placebo group which suggests that, similarly to other treatment modalities for M-TMD as mentioned above, the placebo effect likely played a role in this study.

5. Conclusions

In conclusion, provided the safety, convenience, and likely potential clinical benefit of ESWT in the treatment of M-TMD suggested by the results of this prospective pilot study, it is worthwhile to explore this application further with well-designed future prospective trials with larger sample sizes.

Author Contributions: Conceptualization, D.T.S.L. and Y.Y.L.; methodology, D.T.S.L.; formal analysis, D.T.S.L. and K.Y.L.; writing—original draft preparation, D.T.S.L.; writing—review and editing, Y.Y.L. All authors have read and agreed to the published version of the manuscript.

Funding: This research received no external funding.

Institutional Review Board Statement: The study was conducted in accordance with the Declaration of Helsinki, and approved by the Institutional Review Board of the University of Hong Kong/Hospital Authority Hong Kong West Cluster (HKU/HA HKW IRB) (IRB Reference Number: UW 20-704; Date of approval: 18 November 2020).

Informed Consent Statement: Informed consent was obtained from all subjects involved in the study.

Data Availability Statement: Data is available in the manuscript.

Conflicts of Interest: The authors declare no conflict of interest.

References

1. Schiffman, E.; Ohrbach, R.; Truelove, E.; Look, J.; Anderson, G.; Goulet, J.P.; List, T.; Svensson, P.; Gonzalez, Y.; Lobbezoo, F.; et al. Diagnostic Criteria for Temporomandibular Disorders (DC/TMD) for Clinical and Research Applications: Recommendations of the International RDC/TMD Consortium Network* and Orofacial Pain Special Interest Groupdagger. *J. Oral Facial Pain Headache* **2014**, *28*, 6–27. [CrossRef] [PubMed]
2. Magnusson, T.; Syren, M. Therapeutic jaw exercises and interocclusal appliance therapy. A comparison between two common treatments of temporomandibular disorders. *Swed. Dent. J.* **1999**, *23*, 27–37. [PubMed]
3. Lövgren, A.; Visscher, C.M.; Alstergren, P.; Lobbezoo, F.; Häggman-Henrikson, B.; Wänman, A. The outcome of a temporomandibular joint compression test for the diagnosis of arthralgia is confounded by concurrent myalgia. *Clin. Oral Investig.* **2020**, *24*, 97–102. [CrossRef] [PubMed]
4. Li, D.T.S.; Leung, Y.Y. Temporomandibular Disorders: Current Concepts and Controversies in Diagnosis and Management. *Diagnostics* **2021**, *11*, 459. [CrossRef] [PubMed]
5. Kuttila, M.; Niemi, P.M.; Kuttila, S.; Alanen, P.; Le Bell, Y. TMD treatment need in relation to age, gender, stress, and diagnostic subgroup. *J. Orofac. Pain* **1998**, *12*, 67–74.
6. Gauer, R.L.; Semidey, M.J. Diagnosis and treatment of temporomandibular disorders. *Am. Fam. Physician* **2015**, *91*, 378–386.
7. Craane, B.; Dijkstra, P.U.; Stappaerts, K.; De Laat, A. One-year evaluation of the effect of physical therapy for masticatory muscle pain: A randomized controlled trial. *Eur. J. Pain* **2012**, *16*, 737–747. [CrossRef]
8. Chan, N.H.Y.; Ip, C.K.; Li, D.T.S.; Leung, Y.Y. Diagnosis and Treatment of Myogenous Temporomandibular Disorders: A Clinical Update. *Diagnostics* **2022**, *12*, 2914. [CrossRef]
9. Shimada, A.; Ishigaki, S.; Matsuka, Y.; Komiyama, O.; Torisu, T.; Oono, Y.; Sato, H.; Naganawa, T.; Mine, A.; Yamazaki, Y.; et al. Effects of exercise therapy on painful temporomandibular disorders. *J. Oral Rehabil.* **2019**, *46*, 475–481. [CrossRef]
10. Heir, G.M. The Efficacy of Pharmacologic Treatment of Temporomandibular Disorders. *Oral Maxillofac. Surg. Clin. N. Am.* **2018**, *30*, 279–285. [CrossRef]
11. Incorvati, C.; Romeo, A.; Fabrizi, A.; Defila, L.; Vanti, C.; Gatto, M.R.A.; Marchetti, C.; Pillastrini, P. Effectiveness of physical therapy in addition to occlusal splint in myogenic temporomandibular disorders: Protocol of a randomised controlled trial. *BMJ Open* **2020**, *10*, e038438. [CrossRef] [PubMed]
12. Kutuk, S.G.; Ozkan, Y.; Kutuk, M.; Ozdas, T. Comparison of the Efficacies of Dry Needling and Botox Methods in the Treatment of Myofascial Pain Syndrome Affecting the Temporomandibular Joint. *J. Craniofac. Surg.* **2019**, *30*, 1556–1559. [CrossRef] [PubMed]
13. Patel, J.; Cardoso, J.A.; Mehta, S. A systematic review of botulinum toxin in the management of patients with temporomandibular disorders and bruxism. *Br. Dent. J.* **2019**, *226*, 667–672. [CrossRef] [PubMed]
14. Nasri-Heir, C.; Epstein, J.B.; Touger-Decker, R.; Benoliel, R. What should we tell patients with painful temporomandibular disorders about what to eat? *J. Am. Dent. Assoc.* **2016**, *147*, 667–671. [CrossRef] [PubMed]
15. Chaussy, C.; Brendel, W.; Schmiedt, E. Extracorporeally induced destruction of kidney stones by shock waves. *Lancet* **1980**, *2*, 1265–1268. [CrossRef] [PubMed]
16. Li, X.; Zhang, L.; Gu, S.; Sun, J.; Qin, Z.; Yue, J.; Zhong, Y.; Ding, N.; Gao, R. Comparative effectiveness of extracorporeal shock wave, ultrasound, low-level laser therapy, noninvasive interactive neurostimulation, and pulsed radiofrequency treatment for treating plantar fasciitis: A systematic review and network meta-analysis. *Medicine* **2018**, *97*, e12819. [CrossRef]
17. Sun, J.; Gao, F.; Wang, Y.; Sun, W.; Jiang, B.; Li, Z. Extracorporeal shock wave therapy is effective in treating chronic plantar fasciitis: A meta-analysis of RCTs. *Medicine* **2017**, *96*, e6621. [CrossRef]
18. Lu, Z.; Lin, G.; Reed-Maldonado, A.; Wang, C.; Lee, Y.C.; Lue, T.F. Low-intensity Extracorporeal Shock Wave Treatment Improves Erectile Function: A Systematic Review and Meta-analysis. *Eur. Urol.* **2017**, *71*, 223–233. [CrossRef]
19. Xiang, J.; Wang, W.; Jiang, W.; Qian, Q. Effects of extracorporeal shock wave therapy on spasticity in post-stroke patients: A systematic review and meta-analysis of randomized controlled trials. *J. Rehabil. Med.* **2018**, *50*, 852–859. [CrossRef]

20. Bannuru, R.R.; Flavin, N.E.; Vaysbrot, E.; Harvey, W.; McAlindon, T. High-energy extracorporeal shock-wave therapy for treating chronic calcific tendinitis of the shoulder: A systematic review. *Ann. Intern. Med.* **2014**, *160*, 542–549. [CrossRef]

21. Dymarek, R.; Halski, T.; Ptaszkowski, K.; Slupska, L.; Rosinczuk, J.; Taradaj, J. Extracorporeal shock wave therapy as an adjunct wound treatment: A systematic review of the literature. *Ostomy Wound Manag.* **2014**, *60*, 26–39.

22. Aguilera-Sáez, J.; Muñoz, P.; Serracanta, J.; Monte, A.; Barret, J.P. Extracorporeal shock wave therapy role in the treatment of burn patients. A systematic literature review. *Burns* **2019**, *7*, 1525–1532. [CrossRef] [PubMed]

23. Schenk, I.; Vesper, M.; Nam, V.C. Initial results using extracorporeal low energy shockwave therapy ESWT in muscle reflex-induced lock jaw. *Mund Kiefer Gesichtschir.* **2002**, *6*, 351–355. [CrossRef] [PubMed]

24. Kim, Y.H.; Bang, J.I.; Son, H.J.; Kim, Y.; Kim, J.H.; Bae, H.; Han, S.J.; Yoon, H.J.; Kim, B.S. Protective effects of extracorporeal shockwave on rat chondrocytes and temporomandibular joint osteoarthritis; preclinical evaluation with in vivo(99m)Tc-HDP SPECT and ex vivo micro-CT. *Osteoarthr. Cartil.* **2019**, *27*, 1692–1701. [CrossRef]

25. Li, W.; Wu, J. Treatment of Temporomandibular Joint Disorders by Ultrashort Wave and Extracorporeal Shock Wave: A Comparative Study. *Med. Sci. Monit.* **2020**, *26*, e923461. [CrossRef]

26. Al-Saleh, M.A.; Alsufyani, N.A.; Saltaji, H.; Jaremko, J.L.; Major, P.W. MRI and CBCT image registration of temporomandibular joint: A systematic review. *J. Otolaryngol. Head Neck Surg.* **2016**, *45*, 30. [CrossRef]

27. Hu, S.; Shao, Z.; Deng, L. Clinical Manifestations, Imaging Features, and Pathogenic/Prognostic Risk Factors for Temporo-mandibular Disorders (TMD): A Case-Control Study Based on Psychogenic Factors of Patients. *Comput. Math. Methods Med.* **2022**, *2022*, 8279357. [CrossRef]

28. Shen, S.; Ye, M.; Wu, M.; Zhou, W.; Xu, S. MRI and DC/TMD Methods Analyze the Diagnostic Accuracy of the Change in Articular Disc of Temporomandibular Joint. *Comput. Math. Methods Med.* **2022**, *2022*, 1770810. [CrossRef]

29. Larheim, T.A.; Abrahamsson, A.K.; Kristensen, M.; Arvidsson, L.Z. Temporomandibular joint diagnostics using CBCT. *Dentomaxillofac. Radiol.* **2015**, *44*, 20140235. [CrossRef]

30. Dawson, A.; Ghafouri, B.; Gerdle, B.; List, T.; Svensson, P.; Ernberg, M. Effects of experimental tooth clenching on pain and intramuscular release of 5-HT and glutamate in patients with myofascial TMD. *Clin. J. Pain* **2015**, *31*, 740–749. [CrossRef]

31. Lavelle, E.D.; Lavelle, W.; Smith, H.S. Myofascial trigger points. *Anesthesiol. Clin.* **2007**, *25*, 841–851. [CrossRef] [PubMed]

32. Agarwal, V.; Gupta, A.; Singh, H.; Kamboj, M.; Popli, H.; Saroha, S. Comparative Efficacy of Platelet-Rich Plasma and Dry Needling for Management of Trigger Points in Masseter Muscle in Myofascial Pain Syndrome Patients: A Randomized Controlled Trial. *J. Oral Facial Pain Headache* **2022**. [CrossRef] [PubMed]

33. Keltz, E.; Bercovich, E.; Militianu, D.; Minerbi, A. Imaging of Myofascial Taut Bands Using Gadolinium-Enhanced Magnetic Resonance Imaging. *Pain Med.* **2022**, *23*, 1336–1337. [CrossRef] [PubMed]

34. Chen, Q.; Wang, H.J.; Gay, R.E.; Thompson, J.M.; Manduca, A.; An, K.N.; Ehman, R.E.; Basford, J.R. Quantification of Myofascial Taut Bands. *Arch. Phys. Med. Rehabil.* **2016**, *97*, 67–73. [CrossRef] [PubMed]

35. Turo, D.; Otto, P.; Hossain, M.; Gebreab, T.; Armstrong, K.; Rosenberger, W.F.; Shao, H.; Shah, J.P.; Gerber, L.H.; Sikdar, S. Novel Use of Ultrasound Elastography to Quantify Muscle Tissue Changes After Dry Needling of Myofascial Trigger Points in Patients With Chronic Myofascial Pain. *J. Ultrasound Med.* **2015**, *34*, 2149–2161. [CrossRef]

36. Sikdar, S.; Shah, J.P.; Gebreab, T.; Yen, R.H.; Gilliams, E.; Danoff, J.; Gerber, L.H. Novel applications of ultrasound technology to visualize and characterize myofascial trigger points and surrounding soft tissue. *Arch. Phys. Med. Rehabil.* **2009**, *90*, 1829–1838. [CrossRef]

37. Diep, D.; Chen, K.J.Q.; Kumbhare, D. Ultrasound-guided interventional procedures for myofascial trigger points: A systematic review. *Reg. Anesth. Pain Med.* **2021**, *46*, 73–80. [CrossRef]

38. Hachisuka, R.; Taguchi, S.; Moriwaki, K.; Oshita, K.; Umeda, A.; Tsutsumi, Y.M. Subacute postoperative myofascial pain diagnosed and treated successfully by ultrasound: A case after laparoscopic hepatectomy. *JA Clin. Rep.* **2022**, *8*, 53. [CrossRef]

39. Von Korff, M.; Ormel, J.; Keefe, F.J.; Dworkin, S.F. Grading the severity of chronic pain. *Pain* **1992**, *50*, 133–149. [CrossRef]

40. Al-Moraissi, E.A.; Conti, P.C.R.; Alyahya, A.; Alkebsi, K.; Elsharkawy, A.; Christidis, N. The hierarchy of different treatments for myogenous temporomandibular disorders: A systematic review and network meta-analysis of randomized clinical trials. *Oral Maxillofac. Surg.* **2022**, *26*, 519–533. [CrossRef]

41. Al-Moraissi, E.A.; Farea, R.; Qasem, K.A.; Al-Wadeai, M.S.; Al-Sabahi, M.E.; Al-Iryani, G.M. Effectiveness of occlusal splint therapy in the management of temporomandibular disorders: Network meta-analysis of randomized controlled trials. *Int. J. Oral Maxillofac. Surg.* **2020**, *49*, 1042–1056. [CrossRef] [PubMed]

42. van Grootel, R.J.; Buchner, R.; Wismeijer, D.; van der Glas, H.W. Towards an optimal therapy strategy for myogenous TMD, physiotherapy compared with occlusal splint therapy in an RCT with therapy-and-patient-specific treatment durations. *BMC Musculoskelet. Disord.* **2017**, *18*, 76. [CrossRef] [PubMed]

43. Michelotti, A.; Parisini, F.; Farella, M.; Cimino, R.; Martina, R. Muscular physiotherapy in patients with temporomandibular disorders. Controlled clinical trial. *Minerva Stomatol.* **2000**, *49*, 541–548. [PubMed]

44. Cioffi, I.; Perrotta, S.; Ammendola, L.; Cimino, R.; Vollaro, S.; Paduano, S.; Michelotti, A. Social impairment of individuals suffering from different types of chronic orofacial pain. *Prog. Orthod.* **2014**, *15*, 27. [CrossRef]

45. Delcanho, R.; Val, M.; Guarda Nardini, L.; Manfredini, D. Botulinum Toxin for Treating Temporomandibular Disorders: What is the Evidence? *J. Oral Facial Pain Headache* **2022**, *36*, 6–20. [CrossRef]

46. Conti, P.C. Low level laser therapy in the treatment of temporomandibular disorders (TMD): A double-blind pilot study. *Cranio* **1997**, *15*, 144–149. [CrossRef]
47. Wu, J.Y.; Zhang, C.; Xu, Y.P.; Yu, Y.Y.; Peng, L.; Leng, W.D.; Niu, Y.M.; Deng, M.H. Acupuncture therapy in the management of the clinical outcomes for temporomandibular disorders: A PRISMA-compliant meta-analysis. *Medicine* **2017**, *96*, e6064. [CrossRef]
48. Ansari, M.; Baradaran Mahdavi, S.; Vahdatpour, B.; Lahijanian, A.; Khosrawi, S. Effects of Dry Needling and Low-Power Laser for the Treatment of Trigger Points in the Upper Trapezius Muscle: A Randomized Clinical Trial. *J. Chiropr. Med.* **2022**, *21*, 288–295. [CrossRef]
49. Al-Moraissi, E.A.; Goddard, G.; Christidis, N. Are acupuncture and dry needling effective in the management of masticatory muscle pain: A network meta-analysis of randomised clinical trials. *J. Oral Rehabil.* **2022**. [CrossRef]

Article

The Role of Temporomandibular Joint Arthroscopy for Diagnosis and Surgical Management of Synovial Chondromatosis

Salvatore Sembronio [1,*], Luca Raccampo [1], Alessandro Tel [1], Michele Di Cosola [2], Stefania Troise [3], Giovanni Dell'Aversana Orabona [3] and Massimo Robiony [1]

[1] Maxillofacial Surgery Department, Academic Hospital of Udine, Department of Medicine, University of Udine, Piazzale S. Maria della Misericordia 1, 33100 Udine, Italy
[2] Department of Clinical and Experimental Medicine, University of Foggia, 71122 Foggia, Italy
[3] Maxillofacial Unit, Department of Neuroscience, Reproductive Sciences and Dentistry, University of Naples Federico II, 80131 Naples, Italy
* Correspondence: salvatore.sembronio@uniud.it

Abstract: Objective: We report the experience of our maxillo-facial surgery unit into the diagnostic and the therapeutic role of arthroscopy of temporomandibular joint (TMJ) synovial chondromatosis (SC). Materials and Methods: A series of sixteen patients with an imaging, arthroscopical, and histological diagnosis of SC treated with arthroscopy was selected. The surgeries were conducted in the Department of Maxillo-facial surgery, Academic hospital of Udine, from January 2016 to December 2022. Medical history, clinical examination, imaging, arthroscopical, and histological characteristics were recorded and then reviewed and discussed. Results: Clinical improvement, both in pain and in maximum incisal opening (MIO), were noticed in whole patients. Histologically, according to Milgram's classification, the sample was fairly homogeneous. Arthroscopic treatment was successful in 87.5% of the patients. Only two cases of SC relapse were registered and were then submitted to open surgery to perform a total sinovectomy. The data collected were used to develop an SC classification proposal based on clinical, radiological (magnetic resonance imaging), arthroscopical, and histopathological characteristics. Conclusions: TMJ arthroscopy must be considered the first line of treatment for SC, leaving open surgery to relapses cases and those cases with extraarticular extension. A univocal classification is essential to best stage and prognostically characterise this pathology.

Keywords: arthroscopy; temporomandibular surgery; TMJ; synovial chondromatosis

Citation: Sembronio, S.; Raccampo, L.; Tel, A.; Di Cosola, M.; Troise, S.; Dell'Aversana Orabona, G.; Robiony, M. The Role of Temporomandibular Joint Arthroscopy for Diagnosis and Surgical Management of Synovial Chondromatosis. *Diagnostics* **2023**, *13*, 2837. https://doi.org/10.3390/diagnostics13172837

Academic Editor: Luis Eduardo Almeida

Received: 16 July 2023
Revised: 27 August 2023
Accepted: 30 August 2023
Published: 1 September 2023

1. Introduction

Synovial chondromatosis (SC) is a rare and debilitating disorder defined by the World Health Organization as a benign nodular cartilaginous proliferation arising from the joint synovium, bursae, or tendon sheaths [1]. This abnormal growth is characterized by metaplastic changes and by the formation of nodules of highly cellular hyaline cartilage, which may become pedunculated and detach from the synovial membrane, forming loose bodies (LBs) within the joint space [2]. The first mention of SC dates back to 1558, by Ambroise Pare [3], followed in 1764 by Baron Albrecht von Haller, who reported the presence of LBs in the temporomandibular joint (TMJ) [4], before Georg Axhausen provided a more technical case description of TMJ's SC in 1933 [5]. It typically affects the large joints, like the knee, hip, elbow, wrist, ankle, and shoulder, while it is relatively uncommon in the TMJ. When it occurs in the TMJ, it usually affects the upper articular space, which could lead to expansion of the joint space or capsule and intrajoint fluid collection [6]. However, involvement of the inferior joint compartment, and even extraarticular extension to the infratemporal space, the parotid region, or to the middle cranial fossa, has also been described [7,8]. While it can affect individuals of any age, SC is commonly diagnosed in young to middle-aged adults, with a higher prevalence in women; most of the patients are unilaterally affected without side predilection [9,10]. Its pathogenesis remains unclear, but two forms have been described. Primary SC pathogenesis is unknown, while the onset of secondary SC, which is linked to a more passive process, is associated with arthritic or mechanical conditions such as trauma, inflammatory or degenerative arthritis, and other joint diseases [11]. This secondary form is considered more common and less aggressive [12]. Clinical manifestations typically include pain, swelling, clicking, crepitation, facial asymmetry, functional impairment, deviation, and limitation on mouth opening. An accurate differential diagnosis is crucial to differentiate it from other types of temporomandibular disorders (TMDs), especially in the early stages where the signs and symptoms are similar and unspecific. Employing a combination of clinical assessment, imaging techniques such as computed tomography (CT) and magnetic resonance imaging (MRI), and histopathological examination is essential to confirm the presence of abnormal cartilage growth within the joint. Chen et al. [13] recently proposed a three-type classification based on MRI findings: first type: LBs; second type: homogenous mass; third type: mixture of both LBs and homogeneous mass. In 1977, SC was histologically classified into three phases by Milgram [14]: stage 1, or the early stage, involves metaplasia of the synovial membrane without the presence of LBs; stage 2, or the intermediate stage, demonstrates metaplasia of the synovial membrane with the presence of LB; stage 3, or the final stage, shows only LBs without synovial involvement. Stage 3 of SC can also include secondary calcification of the LBs, a condition referred to as Henderson–Jones syndrome [15]. Once diagnosed, treatment options for SC traditionally consist of complete LBs removal and synoviectomy, aiming to alleviate pain, improve joint function, and preserve the structural integrity of the joint. While traditional treatment approach for TMJ SC included open joint surgery or arthrotomy, which often results into significant tissue damage, prolonged recovery periods, and suboptimal outcomes, recent advancements in arthroscopic techniques have revolutionized the management of this condition. In 1989, McCain and de la Rua first reported and described the arthroscopic treatment of TMJ SC and assessed that open surgery indication must be limited to cases where the LBs are over 3 mm in diameter [16]. Arthroscopy provides a direct and magnified view of the TMJ, enabling thorough examination and diagnosis, but also the treatment of SC, exploiting a minimally invasive approach reducing surgical trauma, postoperative pain, and scarring. Additionally, this technique facilitates better preservation of healthy joint structures, and permits the surgeon to reach the medial aspect of the TMJ, which is reported not to be always possible with open surgery [17]. The aim of this article is to report the authors' experience in treating SC with arthroscopy, clarifying its diagnostic and therapeutic role.

2. Materials and Methods

2.1. Patient Population/Study Design

This is a single-institution, retrospective study conducted in the Department of Maxillofacial surgery, Academic hospital of Udine, from January 2016 to December 2022. A total of sixteen patients with an imaging, arthroscopical, and histological diagnosis of SC were selected. Age, gender, medical history, clinical signs and examination, imaging, and histological characteristics were recorded. Patients included in this study had completed a follow-up period of at least six months, otherwise they would have been excluded from the study, as well as patients with incongruous or missing clinical documentation. No other inclusion/exclusion criteria were established.

2.2. Medical History and Physical Examination

Complete medical history of the patients was collected, mainly focusing on systemic arthritic disease, previous trauma, or TMJ surgery. TMJ function was mainly assessed by measuring the maximum incisal opening (MIO), defined as the distance between the central incisors when the mouth is fully open, in addition to the assessment of lateral and protrusive movements. Moreover, the presence of swelling, functional impairment, clicking, crepitation, facial asymmetry, and deviation on mouth opening was evaluated. Patients were asked to assess preauricular pain using a visual analogue scale (VAS).

2.3. Imaging

All the patients included in the study performed a preoperatory imaging evaluation. An MRI, both T1-weighted and T2-weighted, in mouth closed, half, and fully open position, was performed by all patients. MRI appearances of TMJ SC like joint effusion, presence of LB, proliferative synovium, expanded joint capsule, fluid accumulation within the joint space, and extraarticular involvement were assessed. Those MRI findings were used by two of the authors (S.S. and L.R.) to classify those patients following the comprehensive classification proposal later presented in the discussion section (Table 1). Some patients also performed a CT scan, both in closed and open-mouth position, permitting us to evaluate the presence of calcified LBs and their possibility to shift location, as well as irregularity of joint surfaces, sclerosis, and hyperostosis of the glenoid fossa and mandibular condyle.

Table 1. The comprehensive clinical, radiological, arthroscopical, and histopathological classification proposal presented here. It includes three stages of increasing severity, of which elements could overlap, but most of the matching elements establish the stage. It aims to have a diagnostic but also a prognostic role.

Stage	Clinical Evaluation	MRI Appearance	Arthroscopic Inspection	Histopathological Findings (Milgram)
1	Painless or with occasional pain; maybe present some articular sound; MMO in normal range.	Effusion with no LB; proliferative synovia; no bony alterations.	Normal osseus contours; synovitis; hyperplasia of the synovia; no LB or nodules.	Involves metaplasia of the synovial membrane without the presence of LBs.
2	Patient with frequent pain; joint tenderness; articular sound; reduction of the MMO (35–25 mm).	Effusion with LB; proliferative synovia; initial bony alterations; fluid accumulation within the joint space.	Normal osseus contours; hyperplasia of the synovia with subsynovial nodules; nodules about to detach and LBs; synovial polyp; synovitis.	Demonstrates metaplasia of the synovial membrane with the presence of LBs.
3	Patient with chronic pain sometimes of various entities, headache, joint tenderness; reduction of the MMO (>25 mm).	LB; proliferative synovia; bony alterations and extraarticular involvement; fluid accumulation within the joint space.	Abnormal bone contours and degenerative osseus changes; no articular joint capsule hyperplasia but just LBs; synovial polyp; synovitis, chondromalacia; perforation of the disk; adhesions.	Shows only LBs without synovial involvement.

2.4. Surgical Technique

The arthroscopic treatment of TMJ chondromatosis encompasses various procedures depending on the extent of the disease and the individual patient's needs. These include removal of LB and treatment of synovia (scarification, removing of hyperplastic synovia). All patients were treated under general anaesthesia with nasal intubation. The same surgeon performed all surgeries (S.S.). The Henke-Sass Wolf (Tuttlingen, DE) arthroscopic system (1.9 mm, 0°) was used. TMJ was identified by palpation by opening and closing the patient's mouth. A 19 G needle was introduced in the upper compartment and saline was injected enlarging the upper joint space through a pumping technique. A small incision on the injection point was performed with a No. 11 scalpel blade. At this point, the needle was removed and the trocar with the arthroscopic sheath was inserted into the posterior recess of the upper joint space. The trocar was then removed, and the arthroscope was inserted into the arthroscopic sheath, providing a clear view of the TMJ upper compartment and of the signs of SC such as osseus contours, hyperplasia of the synovia, subsynovial nodules, nodules, LBs, synovial polyp, synovitis, chondromalacia, perforation of the disk, and adhesions (Figures 1–3). This first port acted also as irrigation port, and saline was used to continuously wash out the joint space, removing any debris or remaining LBs. Using a triangulation technique, a second cannula of 2.0 mm was introduced in the anterior recess of the superior joint space, and it was used for instrument passage, drainage, and LBs evacuation (Figure 4). However, changing to a larger cannula system, like a 3.0 mm system, which may provide adequate clearance for removal of large LB, was also performed. Forceps were used to remove the LBs larger than the cannula diameter (Figure 5). LBs were also fragmentated using a cold ablation (coblation) radiofrequency device (COBLATOR™ II Surgery System, Smith & Nephew, UK) (Figures 6 and 7). Coblation is a process that uses a radiofrequency electrical energy passing through saline solution, producing plasma that can be applied precisely to tissues to break molecular bonds within cells. This device was also used to remove the hyperplastic synovia and perform a selective synovectomy of the metaplasic areas. The nodules attached to the subsynovial connective tissue were also precisely coblated. Moreover, coblation provides the possibility to split large LBs (>3 mm) in order to ease their washout or removal. Specimens from the affected synovia and LBs were harvested and sent to the Pathology Department to provide a definitive diagnosis (Figure 8). Manual manipulation of the mandible was performed during the approaches and through the whole surgery to reach all joint zones. At the end of the procedure, an intrarticular injection of 1 cc of hyaluronic acid was performed to ease articular mobilization and for anti-inflammatory purpose, then the arthroscopic sheath and the cannula were removed. Therefore, incisions were closed with sutures.

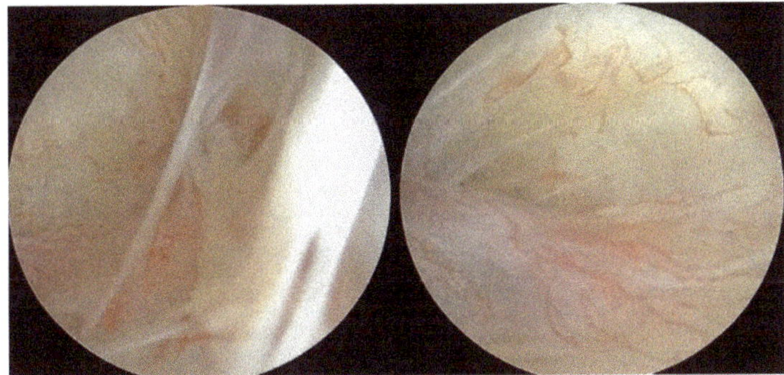

Figure 1. Arthroscopic view of adhesions and moderate to severe synovitis.

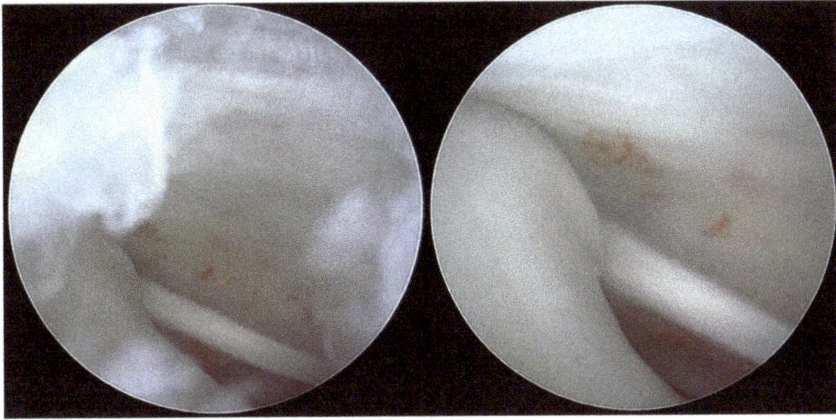

Figure 2. Disk perforation with condyle head exposure and exploration of inferior articular compartment.

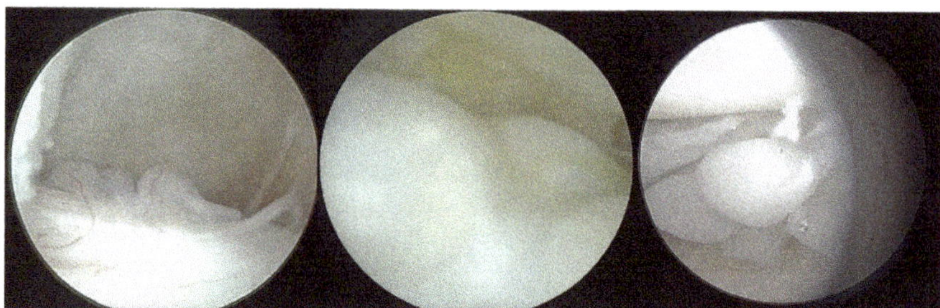

Figure 3. Polyps with vascular axial supply, subsynovial nodules, and LBs.

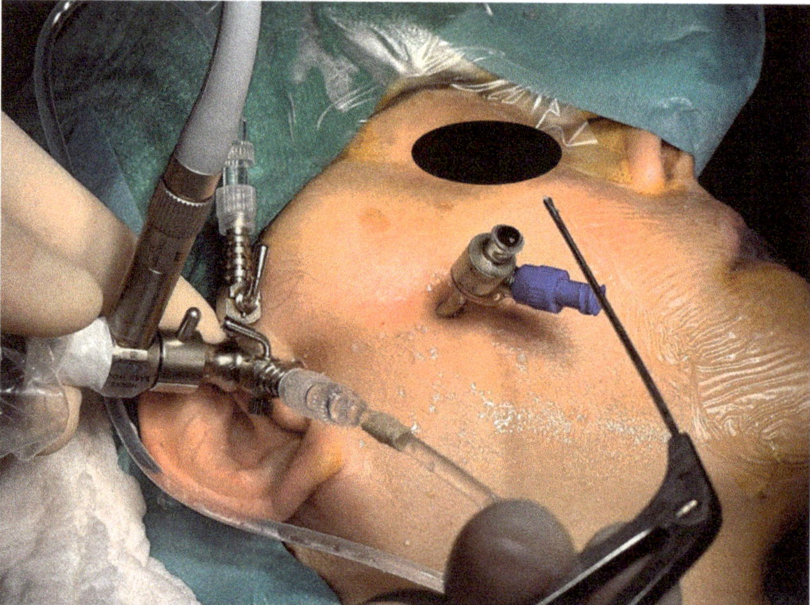

Figure 4. Triangulation technique and LBs washout from the second cannula, which can be seen lying on the patient's face.

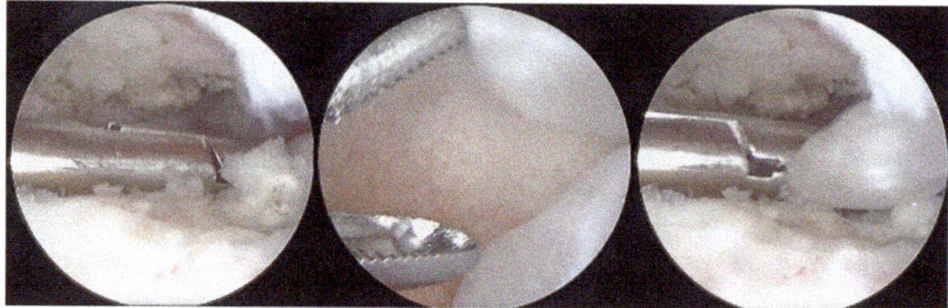

Figure 5. Arthroscopic removal of LBs and synovial nodules using forceps.

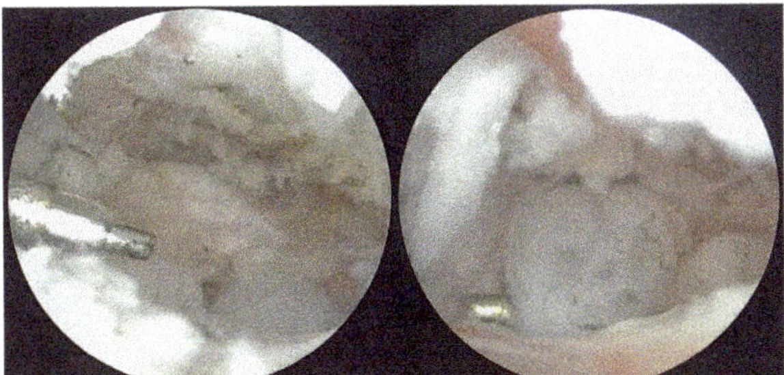

Figure 6. Coblation of metaplasic nodules and division of large LBs to ease their washout.

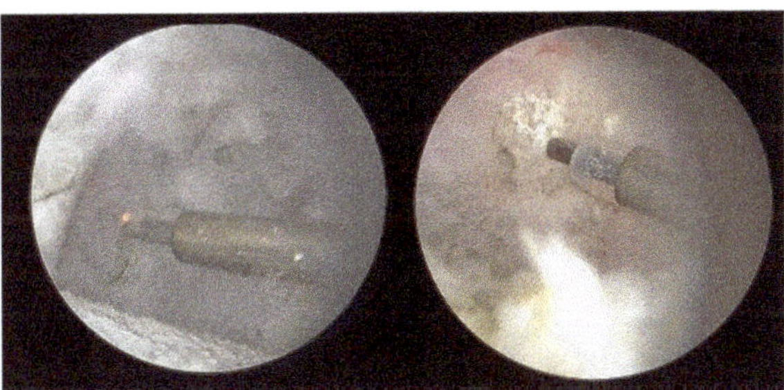

Figure 7. Coblation treatment of the metaplasic synovia, performing the so-called selective synoviectomy.

Figure 8. LB sample sent for histopathological analysis.

2.5. Histological Examination

The LBs and the affected synovia arthroscopically harvested specimens were sent to the Pathology Department of our hospital for histopathological examination. A histopathological diagnosis of SC was made for all patients. Milgram classification (Table 2) was used to stage all the patients.

Table 2. Milgram's histopathological classification published in 1977.

Stage	Histopathological Findings (Milgram)
1	Involves metaplasia of the synovial membrane without the presence of LBs.
2	Demonstrates metaplasia of the synovial membrane with the presence of LBs.
3	Shows only LBs without synovial involvement.

2.6. Follow-Up and Outcome Evaluation

After the arthroscopic procedure, the patients were monitored in a recovery area until they were awake and stable. Pain medications and anti-inflammatory drugs were prescribed to manage postoperative pain. The patients were discharged 1 day after the surgery and follow-up clinical evaluations were scheduled 1 week, 1 month, 3 months, 6 months, and 12 months after surgery. If no problems were detected, the patients were then scheduled for an annual follow-up visit. During the first six months after surgery, all patients underwent articular physiotherapy with mandibular manipulation and joint mobilization. Clinical signs were assessed and recorded. Specifically, two parameters, such as the MIO and the VAS, were used to clinically assess patient outcome. A follow-up MRI was performed on all patients six months after surgery.

3. Results

Between January 2016 and December 2022, 16 patients with clinical and radiological suspicion of SC underwent diagnostic and operative TMJ arthroscopy. Preoperative and postoperative clinical and imaging data (Figure 9) were recorded and are shown in Table 3. Physical examination data at the 6-month follow-up evaluation were analysed to better match the imaging data collected from the 6-month postoperative MRI (Figure 10). Patient age ranged from 24 to 72 years, with a mean of 50.9 years. A sex predominance was observed, with 13 female (81.2%) and just 3 male patients (18.8%). All case reported were monoarticular, with a small prevalence in affected joint of the left side (62.5%) compared to the right side (37.5%). Physical examination records showed a preoperative MIO ranging from 20 to 38 mm with a median of 29.6 mm, while the postoperative MIO ranged from 25 to 48 with a median of 37.9 mm, showing an 8 mm MIO improvement after surgery (Figure 11). Pain assessed with VAS showed a significant improvement of almost 5 units, with the preoperative values ranging from 6 to 8 with a median of 7.1 and the VAS assessed six months after surgery ranging from 0 to 7 with a median of 2.6. All 16 patients were submitted to an MRI before the surgery, while a CT scan was performed in just five patients. The diagnosis of SC was confirmed by histopathological examination and classified following Milgram classification: six patients (37.5%) were classified as a stage 1, as many as the ones classified as stage 2, while four patients (25%) were classified as stage 3. There were no extracapsular soft-tissue involvements that were proven by pathology and MRI in all subjects. It was not possible from the data collected to determine with certainty which form of SC, primary or secondary, the patients were suffering from. The follow-up period ranged from 8 to 71, months with a mean value of 30.6 months. Only 2 of the 16 patients (12.5%) showed a recurrence of SC, and in both cases, this was noticed at the 6-month follow-up evaluation. These two patients were then submitted to open surgery within 6 months of the relapse diagnosis. One of these two patients needed a second arthrotomy 18 months after the first one, because another relapse of SC was noticed at the follow-up MRI.

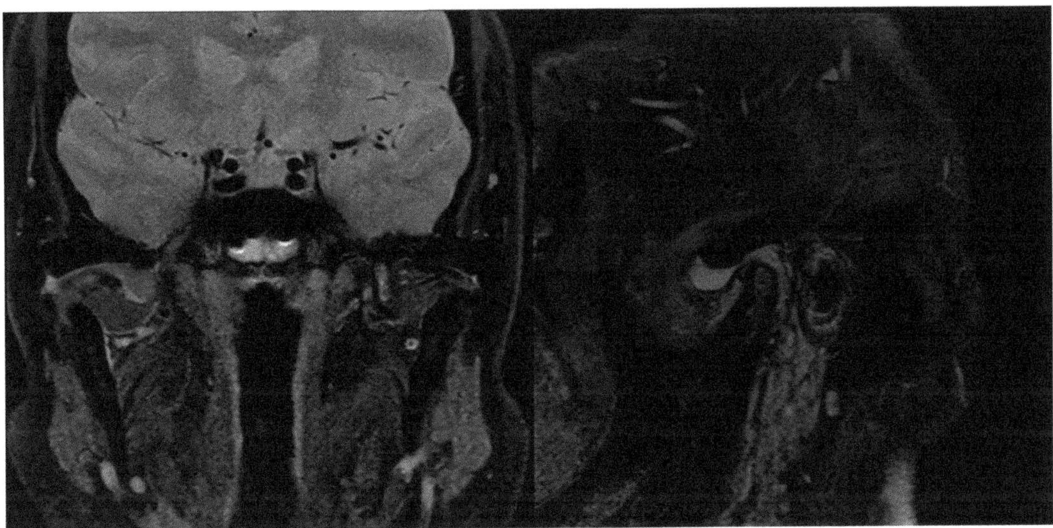

Figure 9. Preoperative sagittal and coronal MRI scans showing, on the right, TMJ joint effusion and LBs, but no bony involvement.

Table 3. Table of our study data. T1 represents a preoperative statement, while T2 represents the 6-month postoperative time.

ID	Sex	Date of Birth	Age (Years)	Surgery Year	Affected TMJ	MIO T1 (mm)	MIO T2 (mm)	VAS T1	VAS T2	MRI	CT	Milgram	Relapse	Follow-Up Period (Months)	Conversion to Open Surgery
1	F	1962	55	2017	Left	35	37	7	3	X	NO	2	NO	71	NO
2	M	1952	66	2018	Right	35	36	6	3	X	NO	1	NO	65	NO
3	F	1969	49	2018	Left	20	25	8	4	X	X	1	YES	60	YES
4	F	1970	49	2019	Left	30	43	6	1	X	NO	2	NO	47	NO
5	F	1952	67	2019	Right	38	48	6	2	X	NO	2	NO	45	NO
6	M	1992	27	2019	Left	30	41	7	1	X	X	2	NO	44	NO
7	F	1979	40	2019	Left	29	45	8	7	X	X	2	YES	43	YES
8	F	1969	52	2021	Right	28	25	8	5	X	X	1	NO	22	NO
9	F	1973	48	2021	Right	28	37	7	0	X	X	1	NO	21	NO
10	F	1954	68	2022	Left	21	45	8	0	X	NO	3	NO	13	NO
11	F	1985	37	2022	Left	25	34	8	3	X	NO	1	NO	12	NO
12	F	1950	72	2022	Right	35	44	7	2	X	NO	3	NO	12	NO
13	F	1998	24	2022	Right	31	38	6	2	X	NO	1	NO	10	NO
14	M	1964	59	2023	Left	37	40	8	1	X	NO	3	NO	9	NO
15	F	1978	45	2023	Left	27	35	7	4	X	NO	2	NO	8	NO
16	F	1966	57	2023	Left	25	34	7	3	X	NO	3	NO	8	NO

Identification (ID); sex: female (F), male (M); maximum incisal opening (MIO); T1 represents a preoperative statement; T2 indicates the 6-month postoperative time; visual analogue scale (VAS); magnetic resonance imaging (MRI); computed tomography (CT).

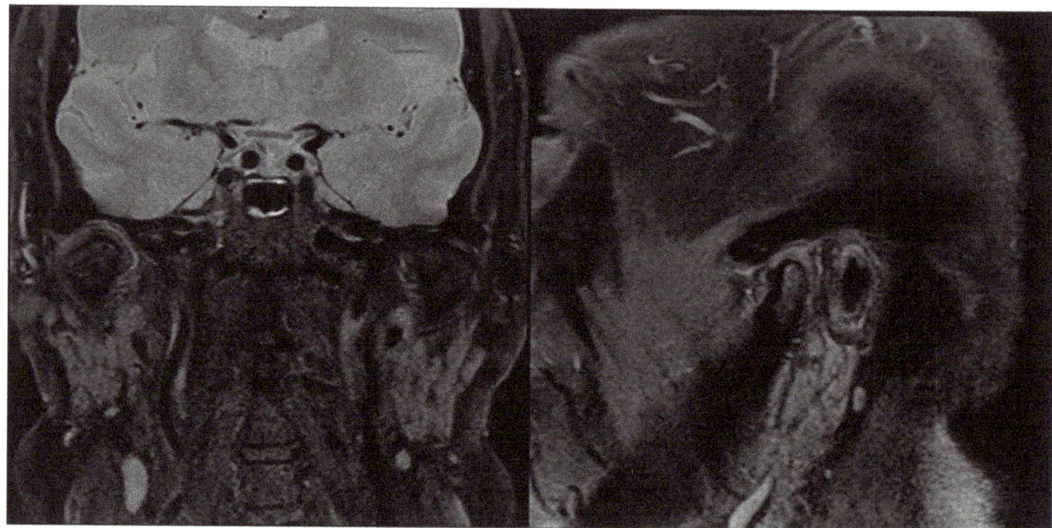

Figure 10. The 6-month postoperative sagittal and coronal MRI scans showing no signs of relapse on the right TMJ.

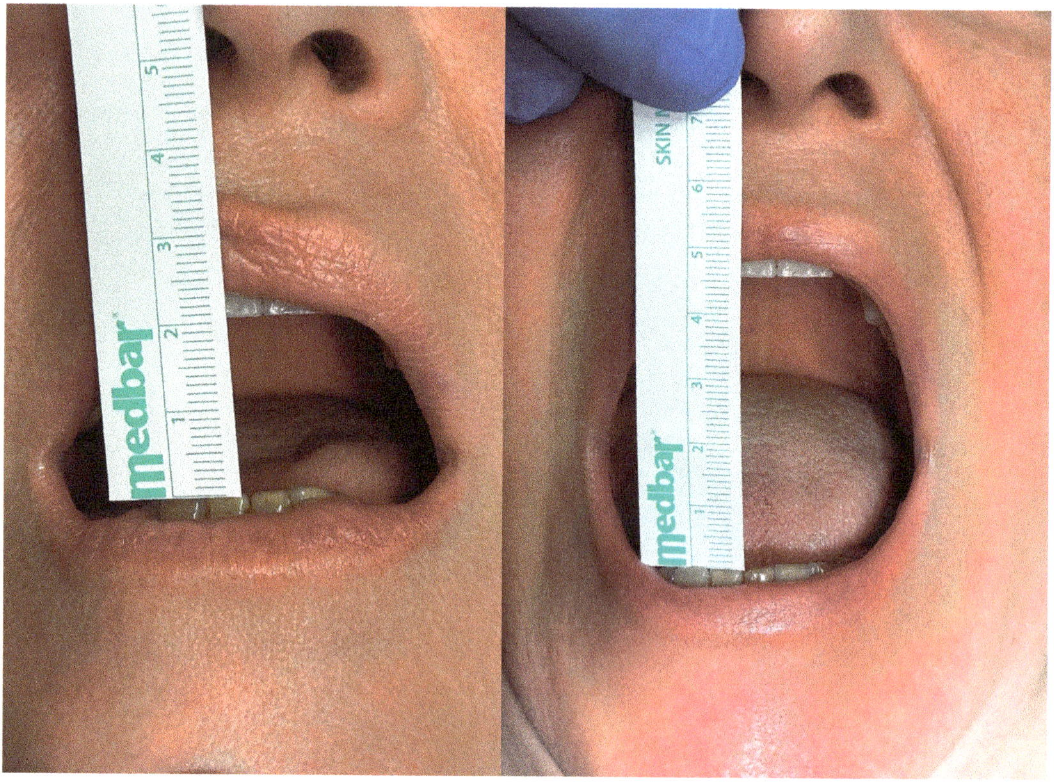

Figure 11. Important visible MIO improvement at 6-month clinical follow-up examination.

4. Discussion

SC was considered a rare condition affecting the TMJ in the past, while recently, reports on the disease have increased. Overviewing the literature, there were mainly reported cases of advanced stage of SC of the TMJ according to Milgram's classification [14]. This could be attributed to the clinical and radiological similarity of SC to other TMDs, which could lead to initial misdiagnosis. We instead reported just three cases of Milgram's stage 3. However, a trend of progressive reduction in the time between occurrence of the first symptoms and provisional diagnosis of SC was recently addressed [18]. A role in this tendency could be represented by the MRI evolution, because this imaging technique is the only one which permits the visualization of the radiological signs of Milgram stage 1. In fact, the absence of LBs does not exclude SC, as the radiographic demonstration of LBs depends on the extent of calcification, and MRI allows the identification of cases not identified on CT, as it can visualise LBs in the early stages [9]. The calcified LBs are often seen in MRI as low and iso-intensity signal nodules of both small round and punctuate forms [18]. Moreover MRI, with its superior contrast resolution, is useful for showing the extension and boundaries of the lesion, assessing internal derangement of the TMJ, and confirming synovial origin of the lesion [19]. MRI SC features mainly include joint effusion, which is best noticed on T2-weighted sequences, LB within the joint space, proliferative synovium, expanded joint capsule, and anterior displacement of the mandibular condyle [9,13,18,19]. Another advantage of MRI over CT is the early detection of extraarticular extension. The MRI is not only useful for the diagnosis of SC, but it also represents the gold standard for follow-up after surgery, which we used to highlight signs of relapse in 2 of the 16 patients. Moreover, since the treatment is surgical, a thorough radiological evaluation is essential to choose the best type of surgery. SC generally occurs in the superior joint space, which could be due to the fact that the superior compartment is larger than the inferior one, and therefore its capacity to produce LBs is greater than that of the inferior compartment, as some have suggested [20,21]. This predilection also makes arthroscopic management of SC feasible as a diagnostic, but especially operative, technique [12]. In view of the various elements that characterise the pathology, and the lack of n univocal classification, the authors propose a classification that encompasses the clinical, radiological MRI, arthroscopical, and histopathological features of SC, the latter expressed by Milgram's classification (Table 1). This follows the principles of the classification of TMDs according to Wilkes [22]. This classification allows and simplifies a complete staging by also providing a prognostic point of view of the disease but obviously needs to be validated. The aim of this proposal is to stimulate experts of the field to assess an unambiguous classification that can best describe the pathology, considering the arthroscopic point of view as well as the histological and radiological ones. The accepted treatment of SC consists of the complete removal of the LBs and synovectomy of the affected synovia. This was historically achieved by open arthrotomy, while in recent years, arthroscopy has been reported to be almost equally effective in selected patients, moving away from its sole diagnostic role [23]. In the past, it was generally stated that open surgery was required if the loose body is over 3 mm in diameter. This is no longer true, thanks to the advancements into the development of arthroscopic instruments which permit us to fragment the LBs [24]. Arthroscopy can also show areas of metaplasic changes of the synovia, which can be coblated with radiofrequency devices. It also permits synovial biopsies to confirm the pathology. Cai et al. [25] previously reported their experience in the arthroscopic treatment of 33 patients affected by SC and broadened the indications based on MRI diagnosis. We have previously stated the importance of arthroscopy even as a complementary approach to open surgery [26], due also to the fact that this technique best permits us to reach the medial aspect of the joint. In aggressive cases with extraarticular extension, or when an involvement of the lower compartment is highlighted, open surgery remains the therapeutic modality of choice. It is usually performed by a preauricular approach, with removal of loose bodies and complete synovectomy. The trend should be that the stage of the disease should guide the choice of the surgical approach. Recurrence is rare in SC but is more frequent in primary SC and in cases of extraarticular extension [27]. Arthroscopy showed a slightly higher rate of relapse than open arthrotomy [28]. The idea is based on the fact that remnants of the synovium may become a source of recurrence, while others state that the condition has a self-limiting character, so that total synovectomy may be unnecessary [29]. As stated before, we reported a recurrence rate of 12.5%, which occurred in Milgram stage 1 and stage 2 patients; thus, in whom the stage indicates a high level of metaplastic proliferation. Those relapses were detected within 6 months from the arthroscopy, making this period adequate for the assessment of recurrence. One of the patients showed a relapse even after the arthrotomy was performed and needed a second open surgery. The patient was classified as a Milgram stage 2. This could indicate that in more proliferative stages even an open arthrotomy may not be sufficient to perform a complete synovectomy. Another possibility is that in the first arthrotomy some remnants of the affected synovia were left in the TMJ. Arthroscopy showed a significant impact on clinical improvement. Both pain, assessed with VAS, and MIO were greatly improved. This corroborates the hypothesis that arthroscopy may represent a more than feasible option to treat this pathology.

5. Conclusions

Understanding this disorder is crucial for clinicians as it allows for early detection, effective management, and improved quality of life for those affected by this condition. Certainly, arthroscopy has a central role in diagnosis of TMJ SC. Most of all, arthroscopy represents a valid option for the treatment of SC. It must be considered the first line of treatment for SC, leaving open surgery to relapses cases and those cases with extraarticular extension. Considering other diagnostic clinical values and increasing the research population could be essential to assess and certify the role of arthroscopy in TMJ SC. Moreover, also in view of the classification proposed here, it is essential to best classify the pathology. Collaboration between clinicians and researchers is crucial in developing standardized protocols and guidelines to ensure the widespread adoption of arthroscopic treatment as the gold standard in managing TMJ SC.

Author Contributions: Conceptualization, S.S. and L.R.; Methodology, S.S. and L.R.; Validation, S.T. and M.R.; Investigation, L.R.; Resources, L.R. and S.T.; Data curation, L.R.; Writing—original draft, L.R.; Writing—review & editing, S.S.; Visualization, A.T., M.D.C., S.T. and G.D.O.; Supervision, G.D.O. and M.R.; Project administration, M.R. All authors have read and agreed to the published version of the manuscript.

Funding: This research received no external funding.

Institutional Review Board Statement: The study was conducted in accordance with the Declaration of Helsinki, and approved by our institute's local independent Institutional Review Board (ID: IRB_45_2020).

Informed Consent Statement: A written informed statement consent to report and publish their cases were obtained from the patients by the authors. All personal data have been obfuscated, and no specific reference can be traced back to identify the patients.

Data Availability Statement: Not applicable.

Conflicts of Interest: The authors declare no conflict of interest.

References

1. Miller, M.V.; King, A.; Mertens, F. Synovial chondromatosis. In *World Health Organization Classification of Tumours. Pathology and Genetics of Tumours of Soft Tissue and Bone*; Fletcher, C.D.M., Unni, K.K., Mertens, F., Eds.; IARC Press: Lyon, France, 2002; p. 246.
2. Mendonca-Caridad, J.J.; Schwartz, H.C. Synovial chondromatosis of the temporomandibular joint: Arthroscopic diagnosis and treatment of a case. *J. Oral Maxillofac. Surg.* **1994**, *52*, 624–625. [CrossRef]
3. Barwell, R. Clinical Lectures on Movable Bodies in Joints. *Br. Med. J.* **1876**, *1*, 403–405. [CrossRef]
4. Haller, A. Elementa Physiologiae Corporus Humani. In *Sumptibus Societatis Typographicae, Bernae*; Elzevirii: Amsterdam, The Netherlands, 1764; p. 8.
5. Axhausen, G. Pathologie und Therapie des Kiefergelenks. *Fortschr. Zahnheilk* **1933**, *9*, 171.

6. Peyrot, H.; Montoriol, P.; Beziat, J.; Barthelemy, I. Synovial chondromatosis of the temporomandibular joint: CT and MRI findings. *Diagn. Interv. Imaging* **2014**, *95*, 613–614. [CrossRef]
7. Xu, M.H.; Ma, X.C.; Guo, C.B.; Yi, B.; Bao, S.D. Synovial chondromatosis of the temporomandibular joint with middle cranial fossa extension. *Int. J. Oral Maxillofac. Surg.* **2007**, *36*, 652–657. [CrossRef]
8. Gil-Salu, J.L.; Lazaro, R.; Aldasoro, J.; Gonzalez-Darder, J.M. Giant solitary synovial chondromatosis of the temporoman-dibular joint with intracranial extension. *Skull Base Surg.* **1998**, *8*, 99–104. [CrossRef]
9. Testaverde, L.; Perrone, A.; Caporali, L.; Ermini, A.; Izzo, L.; D'angeli, I.; Impara, L.; Mazza, D.; Izzo, P.; Marini, M. CT and MR findings in synovial chondromatosis of the temporo-mandibular joint: Our experience and review of literature. *Eur. J. Radiol.* **2011**, *78*, 414–418. [CrossRef] [PubMed]
10. Machado, G.G.; Zambon, C.E.; Lima, J.M.d.S.d.; Paiva, G.L.A.; Martins, V.A.d.O.; Peres, M.P.S.d.M. Synovial chondromatosis of the temporomandibular joint—Clinical, surgical, and imaging findings of a Milgram stage III case series. *Oral Surg. Oral Med. Oral Pathol. Oral Radiol.* **2023**, *135*, e94–e101. [CrossRef] [PubMed]
11. Petito, A.R.; Bennett, J.; Assael, L.A.; Carlotti, A.E. Synovial chondromatosis of the temporomandibular joint: Varying presentation in 4 cases. *Oral Surg. Oral Med. Oral Pathol. Oral Radiol. Endodontol.* **2000**, *90*, 758–764. [CrossRef] [PubMed]
12. Brabyn, P.J.; Capote, A.; Muñoz-Guerra, M.F.; Zylberberg, I.; Rodríguez-Campo, F.J.; Naval-Gías, L. Arthroscopic Management of Synovial Chondromatosis of the Temporomandibular Joint. Case Series and Systematic Review. *J. Maxillofac. Oral Surg.* **2018**, *17*, 401–409. [CrossRef] [PubMed]
13. Chen, M.-J.; Yang, C.; Qiu, Y.-T.; Jiang, Q.; Shi, H.-M.; Wei, W.-B. Synovial chondromatosis of the tempromandibular joint: Relationship between MRI information and potential aggressive behavior. *J. Cranio-Maxillofac. Surg.* **2015**, *43*, 349–354. [CrossRef] [PubMed]
14. Milgram, J.W. Synovial chondromatosis: A histopathologic study of thirty cases. *J. Bone Jt. Surg.* **1997**, *59*, 792–801. [CrossRef]
15. Khanna, J.; Ramaswami, R. Synovial chondromatosis of the temporomandibular joint with intracranial extension—Report of two cases. *Int. J. Oral Maxillofac. Surg.* **2017**, *46*, 1579–1583. [CrossRef] [PubMed]
16. McCain, J.P.; de la Rua, H. Arthroscopic observation and treatment of synovial chondromatosis of the temporomandibular joint: Report of a case and review of the literature. *Int. J. Oral Maxillofac. Surg.* **1989**, *18*, 233–236. [CrossRef]
17. García, L.D.; Medina, B.G.; Rico, M.-S. Synovial chondromatosis of the temporomandibular joint. Technical details of arthroscopic management. *Br. J. Oral Maxillofac. Surg.* **2020**, *59*, 1102–1103. [CrossRef] [PubMed]
18. Wang, P.; Tian, Z.; Yang, J.; Yu, Q. Synovial chondromatosis of the temporomandibular joint: MRI findings with pathological comparison. *Dentomaxillofac. Radiol.* **2012**, *41*, 110–116. [CrossRef] [PubMed]
19. Jia, M.; Xu, Y.; Shao, B.; Guo, Z.; Hu, L.; Pataer, P.; Abass, K.; Ling, B.; Gong, Z. Diagnostic magnetic resonance imaging in synovial chondromatosis of the temporomandibular joint. *Br. J. Oral Maxillofac. Surg.* **2022**, *60*, 140–144. [CrossRef]
20. Sato, J.; Notani, K.-I.; Goto, J.; Shindoh, M.; Kitagawa, Y. Synovial chondromatosis of the temporomandibular joint accompanied by loose bodies in both the superior and inferior joint compartments: Case report. *Int. J. Oral Maxillofac. Surg.* **2010**, *39*, 86–88. [CrossRef]
21. Chen, M.-J.; Yang, C.; Cai, X.-Y.; Jiang, B.; Qiu, Y.-T.; Zhang, X.-H. Synovial chondromatosis in the inferior compartment of the temporomandibular joint: Different stages with different treatments. *J. Oral Maxillofac. Surg.* **2012**, *70*, e32–e38. [CrossRef]
22. Wilkes, C.H. Structural and functional alterations of the temporomandibular joint. *Northwest Dent.* **1978**, *57*, 287–294.
23. Satyanarayan, P.; Roy, I.D.; Issar, Y.; Tomar, K.; Jakka, S. Comparative Evaluation of Success in Cases of Synovial Chondromatosis of Temporomandibular Joint Treated with Temporomandibular Joint Arthroscopy: A Case Series. *J. Maxillofac. Oral Surg.* **2021**, *21*, 1227–1232. [CrossRef] [PubMed]
24. Jacinto, F.S.; Salvatore, S. Temporomandibular Joint Arthroscopic Surgical Techniques: Electrocoagulation and Myotomy Indications. *Atlas Oral Maxillofac. Surg. Clin.* **2022**, *30*, 165–173. [CrossRef] [PubMed]
25. Cai, X.-Y.; Yang, C.; Chen, M.-J.; Jiang, B.; Zhou, Q.; Jin, J.-M.; Yun, B.; Chen, Z.-Z. Arthroscopic management for synovial chondromatosis of the temporomandibular joint: A retrospective review of 33 cases. *J. Oral Maxillofac. Surg.* **2011**, *70*, 2106–2113. [CrossRef] [PubMed]
26. Sembronio, S.; Albiero, A.M.; Toro, C.; Robiony, M.; Politi, M. Arthroscopy with open surgery for treatment of synovial chondromatosis of the temporomandibular joint. *Br. J. Oral Maxillofac. Surg.* **2008**, *46*, 582–584. [CrossRef] [PubMed]
27. Ardekian, L.; Faquin, W.; Troulis, M.J.; Kaban, L.B.; August, M. Synovial chondromatosis of the temporomandibular joint: Report and analysis of eleven cases. *J. Oral Maxillofac. Surg.* **2005**, *63*, 941–947. [CrossRef]
28. D'souza, B.; Dimitroulis, G. A case of recurrence of synovial chondromatosis of the temporomandibular joint. *Oral Surg. Oral Med. Oral Pathol. Oral Radiol. Endodontol.* **2007**, *104*, e59–e61. [CrossRef]
29. van Ingen, J.; de Man, K.; Bakri, I. CT diagnosis of synovial chondromatosis of the temporomandibular joint. *Br. J. Oral Maxillofac. Surg.* **1990**, *28*, 164–167. [CrossRef]

 *diagnostics*

Article

Evaluation of Rheumatic Diseases Affecting the Temporomandibular Joint: A Cone Beam Computed Tomography Study and Literature Review

Zeliha Merve Semerci * and Sevcihan Günen Yılmaz

Department of Oral and Maxillofacial Radiology, Faculty of Dentistry, Akdeniz University, Antalya 07070, Turkey; dentistsevcihan@hotmail.com
* Correspondence: merveertugrul@outlook.com

Abstract: Introduction: Due to the silent manifestation of temporomandibular joint (TMJ), dentists and rheumatologists may neglect treatment for this joint. Aims: The aim of this study was to investigate the TMJ components in patients with various rheumatic diseases and to compare them with a control group based on cone beam computed tomography (CBCT) images. Materials and Methods: This study comprised an assessment of the CBCT images of 65 patients (130 temporomandibular joints) with various rheumatic diseases (mostly rheumatoid arthritis) affecting the TMJ. Moreover, 65 patients (130 temporomandibular joints) with a similar age and gender distribution were examined as the control group. Pathologies were classified into a total of 12 types for the presence of any osseous changes in the condylar head or articular fossa or for joint space narrowing. Statistical analysis of all data was performed with SPSS version 18. The conformity of continuous variables to a normal distribution was examined by the Kolmogorov–Smirnov test. The Mann–Whitney U test was used to compare the means of two independent groups. The Pearson Chi-square test, Yates correction and Fisher's exact test were used in the analysis of categorical variables. Results: The mean age of the patient and control groups was 50 ± 13 and 48 ± 16, respectively, and no statistically significant difference was found between the patient and control groups in terms of age distribution ($p = 0.123$). Condylar erosion, condylar flattening, subcondylar sclerosis, osteophytes, subcortical cysts, articular eminence resorption and articular eminence flattening rates were found to be statistically significantly higher in the patient group than in the control group ($p < 0.05$). Conclusions: Dentomaxillofacial radiologists should examine the bony components of the TMJ in patients with rheumatic diseases, and a multidisciplinary approach involving a dental specialist and rheumatologist is required.

Keywords: temporomandibular joint; cone beam computed tomography; rheumatic diseases

check for updates

Citation: Semerci, Z.M.; Günen Yılmaz, S. Evaluation of Rheumatic Diseases Affecting the Temporomandibular Joint: A Cone Beam Computed Tomography Study and Literature Review. *Diagnostics* **2024**, *14*, 4. https://doi.org/10.3390/diagnostics14010004

Academic Editor: Luis Eduardo Almeida

Received: 22 November 2023
Revised: 12 December 2023
Accepted: 15 December 2023
Published: 19 December 2023

1. Introduction

The temporomandibular joint (TMJ) is a ginglymoarthrodial joint that comprises the mandibular condyle, temporal bone and articular disc, and it is considered one of the most complex joints in the body [1]. Despite the fact that temporomandibular disorders (TMD) are typically associated with orofacial or dental concerns, we should classify intracapsular TMJ disease as arthropathy. The examination of the TMJ is often overlooked during the clinical evaluation of rheumatic patients, even though TMJ involvement has been reported for several rheumatic diseases [2,3]. Prior to investigating orofacial pain, dental complications and occlusal imbalances, underlying arthropathy should be considered in the presence of significant articular disc damage and signs of inflammation [4]. Likewise, dentists and rheumatologists should evaluate patients with rheumatic disease for TMJ involvement.

Rheumatic diseases refer to a group of conditions with a complex pathophysiology affecting multiple organs. Among these various diseases, inflammatory rheumatic arthritic diseases may show some systemic anomalies along with deformities in the synovial joints. In previous studies, it was mentioned that the TMJ is affected in rheumatic diseases such as rheumatoid arthritis (RA), ankylosing spondylitis (AS), juvenile idiopathic arthritis (JIA), systemic sclerosis (SS)/scleroderma, systemic lupus erythematosus (SLE), psoriatic arthritis (PA) and osteoarthritis (OA) [5].

The first principle in understanding how to manage TMD in rheumatic diseases is simply the conception that TMD may be the appearance of an undiagnosed or advanced rheumatic condition. Patients who experience pain on palpation of the peripheral joints are more likely to have soreness on palpation of the TMJ [5]. Clinical findings include sounds, morning stiffness, pain and limitations particularly in the lateral movement of the TMJ. The presence of changes in the TMJ is often neglected, specifically when treatment is focused on other joints. The treatment and monitoring of the temporomandibular joint is extremely important because of the movement impairment and the complexity of the rheumatic disease [6].

Rheumatic diseases affecting the TMJ present a diagnostic challenge to the dentist in the initial stages of the disease, because the symptoms, radiographic findings and laboratory markers are not unequivocal. Clinical assessment still remains a basic step, but a comprehensive radiological evaluation is necessary to confirm the clinical findings in severe cases. There may be a time delay between the appearance of symptoms and radiographic changes, so the absence of pathology on the radiograph does not exclude a possible disease [7].

Radiographic findings of osseous changes in the TMJ can be seen as osteophyte formation, flattening of the mandibular head, cortical erosion, a decrease in joint space, cortical irregularities, bird beak deformity of the condylar head and subcortical cysts [8]. The diagnostic imaging techniques used for rheumatic diseases of the TMJ include panoramic radiography, lateral radiographs, computed tomography (CT), cone beam computed tomography (CBCT), magnetic resonance imaging (MRI), arthrography and arthrotomography [9].

The osseous components of the joints are usually assessed using panoramic radiography and computed tomography, whereas the soft-tissue components are usually assessed using MRI. Ultrasonography is also particularly used for the assessment of the disk and soft tissues. It has been found that CT is superior to plain radiographs and magnetic resonance in detecting early degenerative changes in the mandibular condylar cartilage. For the radiologic diagnosis of degenerative osseous changes of the mandibular condylar cartilage, the reliability is excellent for CBCT [10,11]. Cone beam CT provides high-resolution multiplanar and three-dimensional (3D) images and delivers a substantially lower radiation dose and a shorter exposure time, and it ensures a lower cost compared to multislice CT. CBCT allows the examination of the TMJ anatomy without superimposition or distortion in all sagittal, coronal and axial slices. This technique is easy to perform, is reproducible and delivers a relatively low dose to the patient [11].

Thus, the objective of the current study was to compare osseous changes of the TMJ on CBCT images between patients with various rheumatic diseases affecting the TMJ and healthy controls.

2. Materials and Methods

This retrospective study was approved by the Research Ethics Board of Akdeniz University in Antalya, Turkey (70904504-66). All study procedures were performed in full compliance with the principles of the Declaration of Helsinki and comparable ethical standards.

The CBCT images of 65 patients (40 females, 25 males) who had various rheumatic diseases and no clinical complaints or signs and symptoms of TMD, based on their medical records, and a control group consisting of 65 age- and gender-matched patients (34 females, 31 males) who also had no clinical complaints or signs and symptoms of TMD or rheumatic disease but required other dental treatment necessitating CBCT radiographic imaging—such as pre-surgical planning for the removal of impacted teeth, dental implant planning, paranasal sinus imaging or other various dental treatments—were included in the study. These patients were referred to the Department of Oral and Maxillofacial Radiology, Faculty of Dentistry of Akdeniz University, between November 2021 and December 2022.

To ensure image standardization and evaluation optimization, only images with a region of interest of 7 × 15 cm were employed. Informed consent was obtained from all individual participants included in the study.

Patients were excluded if they had a previous history of TMJ treatment or surgery, had undergone orthodontic treatment, had a previous history of trauma to the jaws, had TMJ pain associated with orofacial pain disorder, had neurological/neuropathic, endocrine or immune/autoimmune diseases with widespread pain, had a previous history of radiation treatment to the head and neck or there were inadequate existing clinical and CBCT data.

All CBCT images were acquired by the same X-ray technician and using the Veraview X800 (Morita, Kyoto, Japan) with a tube voltage of 100 kV, a tube current of 5 mA, an exposure time of 17.86 s and a voxel size of 0.125 mm.

The region of interest was 7 × 15 cm to cover the TMJ with the inferior orbital margin as the upper limit. During scanning, the patient position was set according to the manufacturer instructions, marked by a laser beam with the machine. Another vertical laser beam was aligned 4 cm in front of the condyle. Patients were instructed to swallow and bite on the bite block. The patient's head was supported as recommended by the manufacturer, using a forehead support and chin rest. In order to evaluate the condylar position in the fossa and joint space, all examinations were performed in a central occlusal relationship according to the technique described in the study of Tsiklakis et al. [12].

A series of axial views of 1-mm thickness was automatically produced following the reconstruction of the raw data. TMJs were evaluated from reconstructed lateral slices perpendicular to the long axis of the condyle, coronal slices parallel to the long axis and central lateral image of the joint. To avoid misinterpretation, osseous changes had to be found in at least two consecutive slices. All images were evaluated by exporting to Digital Imaging and Communications in Medicine (DICOM) format. Images were viewed using a Dell monitor (22″ Full HD 1920 × 1080 display) in a dimly lit room [13,14].

There is no available radiographic scoring method to measure and evaluate TMJ changes in RA, as with Larsen's classification, which is used to evaluate other individual joints; therefore, TMJ changes were evaluated as radiographic osteoarthritic features based on DC/TMD [15,16].

The two observers were asked to evaluate the following imaging characteristics relating to osseous changes of the condyles and in the articular fossa: (1) flattening, defined as a flat osseous contour deviating from the convex form; (2) erosion, defined as an area of decreased density of the cortical bone and the adjacent subcortical bone; (3) osteophytes, defined as marginal hypertrophy with sclerotic borders of the bony tissue arising from the surface of the condyle; (4) subcortical sclerosis, defined as an area of increased density of the cortical plate extending towards the bone marrow; (5) a subchondral cyst, defined as a small cavity underneath the articular surface that diverges from the normal bone marrow; (6) a bifid condyle; (7) loose joint bodies, defined as calcified structures that are discontinuous with the soft tissue or osseous structures of the joint; (8) joint space narrowing, defined as a reduction in space (<1.5 mm) in all anterior, superior and posterior directions; (9) increased joint space, defined as when the distance between the condylar head and mandibular fossa was more than 4 mm; (10) flattening of the articular eminence; (11) resorption in the articular eminence; and (12) ankylosis, defined as bony contact between the mandibular condyle and mandibular fossa [13,14,16].

Statistical Analyses

The statistical analysis of all data was performed using IBM SPSS Statistics 18© Copyright SPSS Inc. Chicago, IL, USA, 1989, 2010. The conformity of continuous variables to a normal distribution was examined by the Kolmogorov–Smirnov test. Categorical variables in the study are presented as a frequency (n) and percentage (%). Continuous variables are presented with mean ± standard deviation (SD) and median (IQR 25–75) values. Since parametric test assumptions were not provided, the Mann–Whitney U test was used to compare the means of two independent groups. The Pearson Chi-square test,

Yates correction and Fisher's exact test were used in the analysis of categorical variables. The agreement between the two observations was examined with the Kappa statistic. The statistical significance level was accepted as 0.05 in the study.

3. Results

The mean age of the sample was 48 years (range: 39–61 years). Table 1 shows the distribution of the sample according to age and gender.

Table 1. Demographic characteristics in patient and control groups.

Variables	Patient Group ($n = 130$)	Control Group ($n = 130$)	Total ($n = 260$)	p
Gender, n (%)				
Male	50 (38.5%)	68 (52.3%)	118 (45.4%)	0.025 *
Female	80 (61.5%)	62 (47.7%)	142 (54.6%)	
Age (years)				
Mean ± SD	50 ± 13	48 ± 16	49 ± 15	0.123 $^\mu$
Median (IQR)	50 (41–61)	47 (34–57)	48 (39–61)	

* Pearson Chi-square test, $^\mu$ Mann–Whitney U test.

A total number of 260 TMJs (130 individuals) were evaluated by two observers blinded to the clinical characteristics of the patients and controls. Age, gender and osseous changes were recorded for each patient. Of the 130 participants, 65 were selected as the patient group and 65 as the age- and gender-matched control group.

The rheumatic disease distribution of the 65 patients in the patient group was as follows: RA 41, osteoarthritis (OA) 8, familial mediterranean fever (FMF) 4, scleroderma 2, Sjögren's syndrome 1, systemic lupus erythematosus (SLE) 1, psoriatic arthritis (PA) 1. Figure 1 shows the distribution of diseases in the patient group.

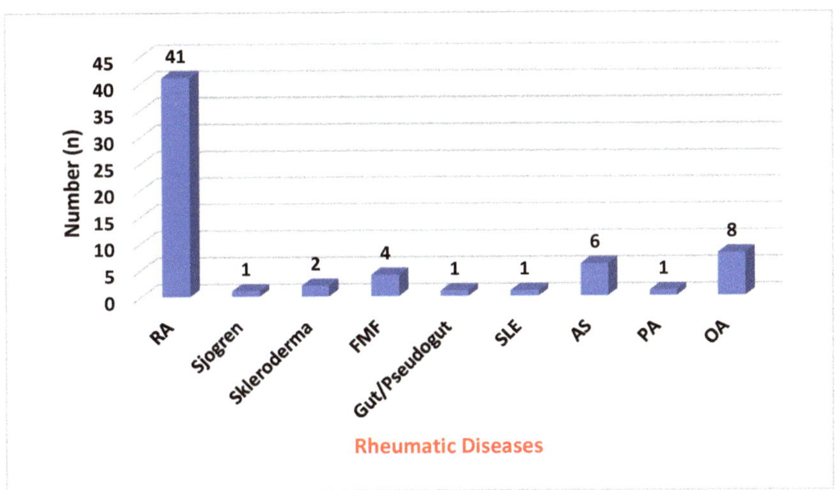

Figure 1. Distribution of diseases in the patient group.

Overall, 45.4% of the patients were male; this rate was 38.5% in the patient group and 52.3% in the control group, and the difference was statistically significant ($p = 0.025$). The median age was calculated as 50 (41–61) years in the patient group, 47 (34–57) years in the control group and 48 (39–61) years in the whole group. In terms of age distribution, the patient and control groups were similar ($p = 0.123$) (Table 1).

Of the 130 TMJs included in the patient group, 69 (53.1%) showed condylar flattening, 54 (41.5%) showed subcondylar sclerosis, 44 (33.8%) showed subcortical cysts, 3 (2.3%) showed bifid condyles, 21 (16.2%) showed articular eminence resorption, 6 (4.6%) showed loose joint bodies, 50 (38.5%) showed osteophytes, 5 (3.8%) showed ankylosis and 25 (19.2%) showed joint space narrowing in both the first and second observer evaluations. In the patient group, an increased joint space was not observed in any TMJs in both observer evaluations.

In accordance with the first observer, condylar flattening was determined in 49 (37.7%) TMJs in the patient group, while the second observer determined condylar flattening in 48 (36.9%) TMJs. Similarly, while the first observer detected articular fossa flattening in 10 (7.7%) TMJs, the second observer detected articular eminence flattening in 11 TMJs (8.5%).

Analyzing the temporomandibular joint osseous changes in the patient and control groups, condylar erosion, condylar flattening, subcondylar sclerosis, osteophytes, subcortical cysts, the resorption of articular eminence and the flattening of articular eminence rates were found to be statistically significantly higher in the patient group compared to the control group, both in the evaluations of the first observer and the second observer ($p < 0.05$). Bifid condyles, loose joint bodies, ankylosis, joint space narrowing and increased joint spaces were found to be statistically similar in both the first and second observer evaluations in the patient and control groups ($p > 0.05$). When the compatibility of the evaluations of the first observer and the second observer was analyzed with the Kappa statistic, it was determined that all evaluations were statistically significantly compatible ($p < 0.001$). A very high level of agreement was found for the values of condylar erosion ($\kappa = 0.97$), condylar flattening ($\kappa = 0.98$), subcondylar sclerosis ($\kappa = 0.99$), osteophytes ($\kappa = 0.99$), flattening of articular eminence ($\kappa = 0.95$) and joint space narrowing ($\kappa = 0.98$). In the evaluations of subcortical cysts, bifid condyles, the resorption of articular eminence, loose joint bodies, ankylosis and an increased joint space, there was complete agreement ($\kappa = 1$) (Table 2).

Table 2. The osseous pathologies of the temporomandibular joint in the patient and control groups and the agreement between observers. Numbers 1 and 2 represent Observer-1 and Observer-2 evaluations, respectively.

Variables	Patient Group (n = 130)	Control Group (n = 130)	Variables	p ¹	Kappa	p ²
Condylar erosion-1	69 (53.1%)	41 (31.5%)	110 (42.3%)	<0.001	0.97	<0.001
Condylar erosion-2	69 (53.1%)	38 (29.2%)	107 (41.2%)	<0.001		
Condylar flattening-1	49 (37.7%)	24 (18.5%)	73 (28.1%)	0.001	0.98	<0.001
Condylar flattening-2	48 (36.9%)	23 (17.7%)	71 (27.3%)	0.001		
Subcortical sclerosis-1	54 (41.5%)	17 (13.1%)	71 (27.3%)	<0.001	0.99	<0.001
Subcortical sclerosis-2	54 (41.5%)	18 (13.8%)	72 (27.7%)	<0.001		
Osteophyte-1	50 (38.5%)	22 (16.9%)	72 (27.7%)	<0.001	0.99	<0.001
Osteophyte-2	50 (38.5%)	21 (16.2%)	71 (27.3%)	<0.001		
Subcortical cyst-1	44 (33.8%)	17 (13.1%)	61 (23.5%)	<0.001	1	<0.001
Subcortical cyst-2	44 (33.8%)	17 (13.1%)	61 (23.5%)	<0.001		
Bifid condyle-1	3 (2.3%)	9 (6.9%)	12 (4.6%)	0.139	1	<0.001
Bifid condyle-2	3 (2.3%)	9 (6.9%)	12 (4.6%)	0.139		
Articular eminence resorption-1	21 (16.2%)	4 (3.1%)	25 (9.6%)	0.001	1	<0.001
Articular eminence resorption-2	21 (16.2%)	4 (3.1%)	25 (9.6%)	0.001		
Articular eminence flattening-1	10 (7.7%)	1 (0.8%)	11 (4.2%)	0.014	0.95	<0.001
Articular eminence flattening-2	11 (8.5%)	1 (0.8%)	12 (4.6%)	0.008		

Table 2. *Cont.*

Variables	Patient Group (*n* = 130)	Control Group (*n* = 130)	Variables	$p^{\,1}$	Kappa	$p^{\,2}$
Loose joint bodies-1	6 (4.6%)	2 (1.5%)	8 (3.1%)	0.281	1	<0.001
Loose joint bodies-2	6 (4.6%)	2 (1.5%)	8 (3.1%)	0.281		
Ankylosis-1	5 (3.8%)	3 (2.3%)	8 (3.1%)	0.722	1	<0.001
Ankylosis-2	5 (3.8%)	3 (2.3%)	8 (3.1%)	0.722		
Joint space narrowing-1	25 (19.2%)	15 (11.5%)	40 (15.4%)	0.122	0.98	<0.001
Joint space narrowing-2	25 (19.2%)	14 (10.8%)	39 (15%)	0.082		
Increased joint space-1	0 (0%)	1 (0.8%)	1 (0.4%)	0.999	1	<0.001
Increased joint space-2	0 (0%)	1 (0.8%)	1 (0.4%)	0.999		

[1] Pearson Chi-square test, Yates correction, Fisher's exact test. [2] Kappa statistic.

Concordance was rated using the criteria of Landis and Koch (1977): 0.01–0.205 slight; 0.21–0.405 acceptable; 0.41–0.605 moderate; 0.61–0.805 considerable; and 0.81–1.005 almost perfect [17].

Exemplary images of osseous changes in the TMJs are shown in Figures 2 and 3.

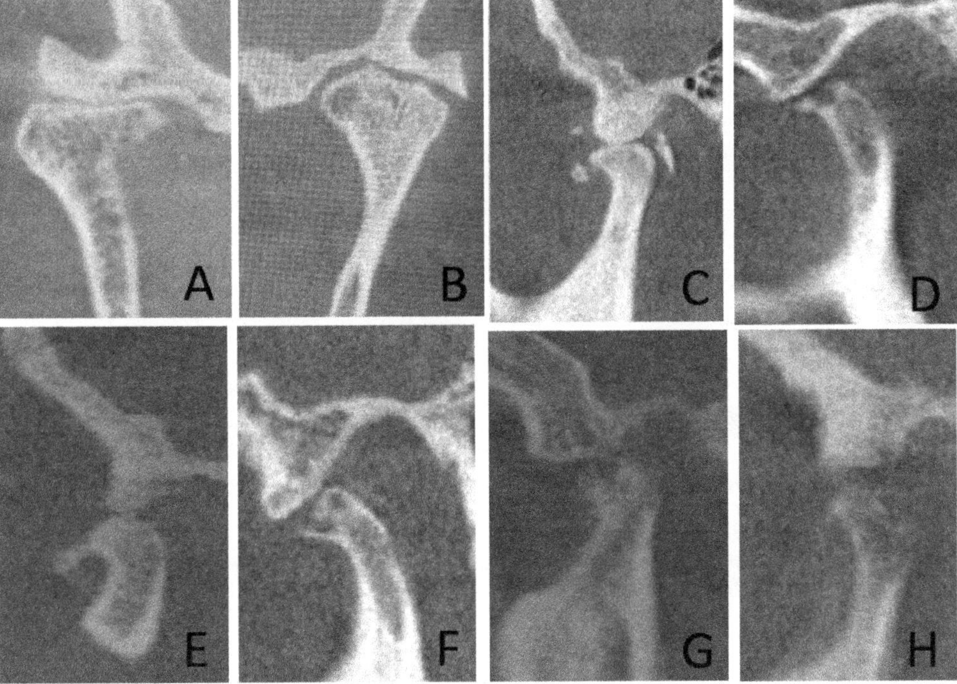

Figure 2. Examples of the CBCT images of osseous changes in the temporomandibular joints (TMJs). (**A**)—flattening of the condyle, subcortical sclerosis and ankylosis in a male with rheumatoid arthritis; (**B**)—joint space narrowing, subcortical cyst formation, condylar surface and articular eminence resorption in a female with scleroderma; (**C**)—loose calcified bodies in a male with rheumatoid arthritis; (**D**)—subcortical cyst formation in a female with psoriatic arthritis; (**E**)—flattening of the condyle and osteophyte formation in a male with osteoarthritis; (**F**)—multiple subcortical cysts in a female with rheumatoid arthritis; (**G**)—progressive condyle erosion in a male with pseudogout; (**H**)—articular eminence and condyle resorption in a female with SLE.

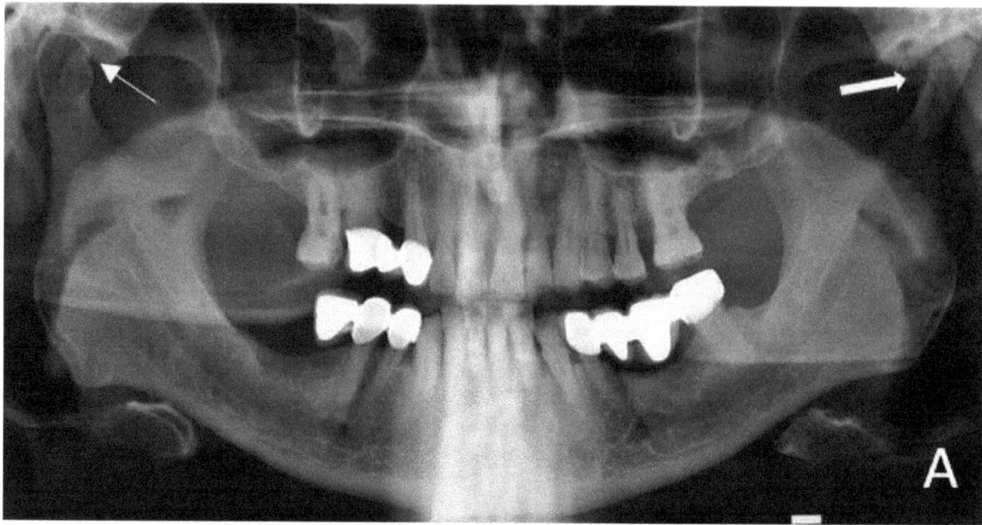

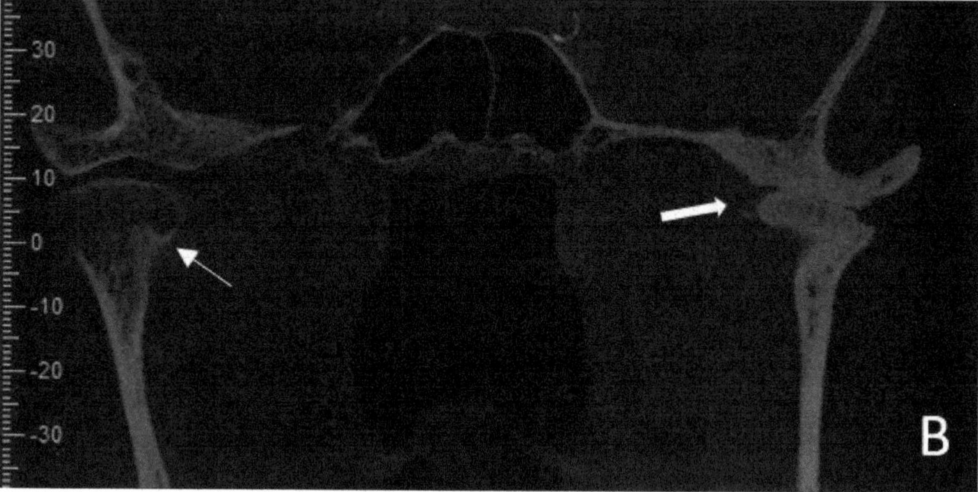

Figure 3. (**A**) presents the panoramic radiographic image of a 62–year–old patient diagnosed with rheumatoid arthritis for the past 12 years. Resorption at the anterior part of the mandibular condyle (indicated by a thin white arrow), multiple radiopacities at the condylar head, condylar flattening and a decrease in the joint space (indicated by a thick white arrow) are observed. (**B**) displays the coronal cone beam computed tomography (CBCT) image of the same patient. Resorption at the anterior part of the mandibular condyle (indicated by a thin white arrow), multiple radiopacities at the condylar head and ankylosis (indicated by a thick white arrow) are observed. The findings are consistent with those observed in panoramic radiography, although the actual extent of the disease can be better determined.

4. Discussion

Rheumatologic diseases can significantly impact the TMJ, causing a range of symptoms and complications. Conditions such as rheumatoid arthritis, systemic lupus erythematosus and psoriatic arthritis are known to affect the TMJ, leading to inflammation, pain and restricted jaw movement [5].

It has been discovered that the prevalence of TMJ involvement in patients with rheumatic disease varies substantially depending on the diagnostic criteria, the population investigated and the methods used to evaluate the TMJ. There are limited studies regarding TMJ radiographic evaluation in patients with rheumatic diseases, especially in PA, systemic scleroderma and AS. Previous studies have focused on the TMJ radiographic findings of RA, while knowledge of the nature of TMJ involvement in other rheumatic diseases is still limited and further research in this field is necessary. Erosions and subcortical cysts of the mandibular condyle are typical initial radiological findings [18].

RA is a systemic, symmetrical, peripheral, inflammatory polyarticular connective tissue disease caused by erosive synovitis, resulting in joint deformity and instability. The prevalence of RA is 1%, affecting women more than men in a 3:1 ratio and an age range between 35 and 45 years [19]. The TMJ is often affected in RA, particularly in its severe form. The incidence of its involvement ranges from 5% to 86% [19]. In addition, a correlation between the laboratory values of various inflammatory markers related to rheumatoid arthritis and the progression of TMD has been reported. Correlations have also been reported to exist between the number of swollen joints, rheumatoid factor (RF), sedimentation rate (ESR), C-reactive protein (CRP), thrombocyte count and plasma tumor necrosis factor alpha levels and temporomandibular joint involvement. According to these studies, TMJ involvement is more prevalent in severe RA patients [20,21]. Ankylosing spondylitis (AS) is a chronic, systemic, inflammatory disease that primarily affects the axial skeleton. TMJ involvement in patients with AS varies between 4% and 35%. Psoriatic arthropathy is an inflammatory seronegative arthritis that affects 5–8% of patients with psoriasis [22]. Half the patients with psoriatic arthropathy have TMJ symptoms, and up to 90% have signs of dysfunction. Involvement of the TMJ is more common and more severe in patients with psoriatic arthropathy than in those with uncomplicated psoriasis and healthy individuals [23]. Systemic sclerosis, also called scleroderma, is a multisystem connective tissue disease with an unknown etiology, and it is defined by infection and fibrotic and vascular changes in the skin and internal organs [24]. Chebbi et al. assessed the effect of systemic sclerosis on TMD. They reported that the early diagnosis of TMD in systemic sclerosis patients is necessary [25]. They observed that the assessment and knowledge of the oral and dental changes due to scleroderma are necessary for dentists because they may lead to the earlier diagnosis of systemic sclerosis [26]. Lupus erythematosus is a chronic autoimmune disease that affects various organs, including joints [27]. Jonsson et al. reported that TMD is common in lupus erythematosus patients [28]. In their study, Crincoli et al. included fifty-five patients diagnosed with systemic lupus erythematosus (SLE), comprising 9 men and 46 women. They compared the temporomandibular disorder (TMD) symptoms in these patients with healthy controls based on the DC/TMD criteria. The results revealed that SLE patients reported a higher frequency (95.8%) of oral and TMJ symptoms, including dysgeusia, stomatodynia, masticatory muscle pain during function, neck and shoulder muscle pain and the presence of tinnitus. These results underscore the heightened prevalence of TMD-related symptoms in individuals with systemic lupus erythematosus, shedding light on the intricate relationship between SLE and temporomandibular joint dysfunction [29].

In the literature, there are studies investigating the impact of rheumatic diseases and inflammatory biomarkers, as well as disease severity, on TMJ involvement [20,30]. In a study conducted by Yılmaz et al., which involved 28 rheumatoid arthritis (RA) patients and 29 control subjects, the progression of TMJ and masticatory muscle involvement was examined using Disease Activity Score 28 (DAS28) scoring, magnetic resonance imaging (MRI) and lateral panoramic radiography. The TMJ symptoms were identified as frequent findings and were associated with the mean duration of the disease in RA. The study suggests that laboratory findings should be taken into consideration when assessing disease-activity-related TMJ involvement [30].

In our study, we observed that the most common temporomandibular joint osseous pathologies in individuals with rheumatic diseases were condylar erosion at a rate of 53%, subcondylar sclerosis at a rate of 41%, osteophytes at a rate of 39% and condylar flattening at a rate of 38%, respectively.

Although the rheumatic disease of the temporomandibular joint can be confused with the findings of a degenerative joint disease due to the advanced age of the patient group, we believe that the effect of this on our results was minimal, as the difference between the findings of the control group and the patient group was statistically significant and our results were consistent with the previous literature [31,32]. Subcortical sclerosis of the condylar surface or fossa is considered a variation, especially with regard to advanced age, remodeling or the association with mandibular hyperfunction as an attempt at adaptation. However, the manifestation of generalized sclerosis of the subchondral bone is associated with joint degradation and may be a result of the presence of TMJ rheumatic disease [33]. Degenerative bone changes, such as the presence of erosion and flattening in the mandibular condyle, are usually noticed in CT scans 5 to 10 years after the onset of symptoms [34]. Helenius et al. examined temporomandibular joint pathologies in patients with various rheumatic diseases, using panoramic and lateral panoramic radiographs, and they observed distinct erosions in 17% of patients with rheumatoid arthritis, 19% of patients with mixed connective tissue disease and 38% of patients with spondyloarthritis. Larheim et al. evaluated 36 patients, 28 of whom were symptomatic, with various rheumatic diseases (mostly rheumatoid arthritis), and found that 25 temporomandibular joints showed TMJ pathology in the CT results [35]. In the study of Wenneberg et al., radiographic changes were found significantly more often in subjects with rheumatoid arthritis (66%), psoriatic arthritis (38%) and ankylosing spondylitis (30%) than in controls (12%). For this reason, they reached the conclusion that rheumatoid arthritis is a more severe disease than psoriatic arthritis or ankylosing arthritis regarding temporomandibular joint involvement [31]. In another study by Wenneberg et al., they used panoramic radiography to compare 90 patients with AS with age- and sex-matched controls. Radiographic changes were observed in 25% of patients and 11% of controls [22]. The first aim of management is to relieve pain. Initial conservative measures include jaw resting, physiotherapy, non-steroidal anti-inflammatory drugs (NSAIDs) and occlusal splints. Around 80% of patients will have their symptoms resolved by conservative treatments alone [36].

Medical specialists that regularly provide treatment for temporomandibular joint disorder in patients without rheumatic conditions are familiar with the predictable pattern of disease advancement that aligns with the Wilkes categorization. Wilkes made a noteworthy observation regarding the significant association between the temporal progression of the biological lesions and various clinical and radiological results, which exhibited a strong correlation [37]. In the context of rheumatic diseases, the temporal development of temporomandibular disorders may exhibit unforeseen patterns. It is important to note that a direct association between clinical observations and radiographic results should not be directly presumed. A multidisciplinary approach is crucial, combining the expertise of rheumatologists, dentists and other healthcare professionals. Rheumatologists play a pivotal role in addressing the systemic aspects of these diseases, managing inflammation and prescribing appropriate medications. Dentists, particularly those specializing in oral and maxillofacial medicine, focus on localized TMJ symptoms, offering treatments such as occlusal splints, physical therapy and, in severe cases, surgical interventions [5]. The collaboration between these disciplines ensures a holistic approach, considering both the systemic and local manifestations of rheumatologic diseases affecting the TMJ, ultimately improving patient outcomes and quality of life.

The clinical examination of TMJ rheumatic disease is insufficient to fully evaluate the osseous and soft-tissue changes in the mandibular condyle, and the need for TMJ imaging is usually determined after a thorough anamnesis and clinical examination. Cone beam computed tomography scans are widely regarded as a valuable imaging method for the viewing of the temporomandibular joint (TMJ) compared to other techniques. Panoramic radiography has several limitations, such as structural distortion, superimposition from the zygomatic process and the inability to show the entire articular surface of the TMJ [38].

Ahmad et al. mentioned in their study that panoramic radiographs also have low reliability and low sensitivity in detecting osseous changes in the TMJ [16].

The lateral radiographs show too much overlap of other anatomical structures, and, in addition, the soft tissues are not visible. Axial tomography provides a sufficient view of the erosions and osteophytes on the surface of the condyle, but their interpretation is difficult. Computed tomography (CT) is valid for the imaging of the mandibular condyle but the device has a high cost and a relatively high radiation dose, and there is poor access to equipment, all of which limits its use for the evaluation of the TMJ [39].

MRI is considered the prime imaging method to evaluate the soft-tissue components of the temporomandibular joint in circumstances when the diagnosis of a soft-tissue pathology is uncertain or when ionizing radiation should be avoided [40]. In conventional MRI, soft tissues typically appear bright on the images, making it challenging to distinguish between different structures, especially when they have similar signal intensities. With advancements in imaging methods, there are studies indicating that MRI can be used to depict the bony components of the TMJ. These methodologies are often known as ultra-short echo time (UTE) and zero echo time (ZTE) sequences. These approaches have demonstrated the capacity to facilitate valuable bony tissue imaging using MRI. However, CBCT or MDCT remains the gold standard for TMJ bony components [29].

Although CBCT is unable to display actual Hounsfield units, which can provide a more valid quantitative assessment of bone density, it has higher sensitivity with regard to viewing the morphology of the osseous components of the joint, cortical bone continuity, subcortical bone destruction and sclerosis. CBCT is better at detecting changes in condylar and articular eminence flattening, osteophyte formation and erosion [41].

CBCT is also superior to CT in analyzing lateral slices in isolation and combining coronal and lateral slices [41]. Therefore, CBCT was the imaging method used in this study.

According to the study of Librizzi et al., in which two different cone beam computed tomography devices were used for the detection of osseous changes in the TMJ, osseous changes in the TMJ could be better differentiated in smaller fields of view (FOV). For this reason, we only included images in 7 × 15 cm FOVs in our study [42].

This study was limited by being a monocenter study and the fact that the CBCT findings were surveyed retrospectively. All patients were referred because of TMD, which included a wide range of symptoms; this may explain the high rate of radiographic degeneration of temporomandibular joint bony structures in the age- and sex-matched control group.

Another limitation of this study was that the clinical evaluation of the patients could not be assessed because it was planned retrospectively. Moreover, there was a limited number of patients with rheumatic diseases other than rheumatoid arthritis; therefore, the group of patients could not be compared among themselves. For future studies, the abovementioned limitations can be overcome with larger patient and control groups, with the aim of helping them to achieve better quality of life.

5. Conclusions

Due to the silent manifestation of TMJ disorders, dentists and rheumatologists may neglect treatment for this joint. A multidisciplinary approach is essential to reduce the physical and psychological consequences of rheumatic diseases affecting the TMJ.

Considering the study results, individuals with rheumatic diseases, even asymptomatic patients, are susceptible to osteoarthritic pathologies in the temporomandibular joint. The assessment of TMJ bony components for osteoarthritic changes in individuals with various rheumatic diseases can be conducted through cone beam computed tomography. To facilitate treatment, early diagnosis and disease prevention, clinicians are advised to employ CBCT imaging when deemed necessary.

Author Contributions: Conceptualization, Z.M.S. and S.G.Y.; methodology, Z.M.S. and S.G.Y.; software, Z.M.S. and S.G.Y.; validation, Z.M.S. and S.G.Y.; formal analysis, Z.M.S. and S.G.Y.; investigation, Z.M.S. and S.G.Y.; resources, Z.M.S. and S.G.Y.; data curation, Z.M.S. and S.G.Y.; writing—original draft preparation, Z.M.S.; writing—review and editing, Z.M.S. and S.G.Y.; visualization, Z.M.S. and

S.G.Y.; supervision, S.G.Y.; project administration, not applicable; funding acquisition, not applicable. All authors have read and agreed to the published version of the manuscript.

Funding: This research received no external funding.

Institutional Review Board Statement: The study was conducted in accordance with the Declaration of Helsinki, and approved by the Institutional Review Board (or Ethics Committee) of Akdeniz University in Antalya, Turkey, 2022 (70904504-66) for studies involving humans.

Informed Consent Statement: Informed consent was obtained from all subjects involved in the study.

Data Availability Statement: Reported results are unavailable due to privacy restrictions.

Conflicts of Interest: The authors declare no conflict of interest.

Abbreviations

TMJ	Temporomandibular joint
CBCT	Cone beam computed tomography
TMD	Temporomandibular disorders
RA	Rheumatoid arthritis
AS	Ankylosing spondylitis
JIA	Juvenile idiopathic arthritis
SS	Systemic sclerosis
SLE	Systemic lupus erythematosus
PA	Psoriatic arthritis
OA	Osteoarthritis
CT	Computed tomography
MRI	Magnetic resonance imaging
RF	Rheumatoid factor
ESR	Sedimentation rate
CRP	C-reactive protein
DICOM	Digital Imaging and Communications in Medicine
MDCT	Multidetector Computed Tomography

References

1. Shaffer, S.M.; Brismee, J.M.; Sizer, P.S.; Courtney, C.A. Temporomandibular disorders. Part 1: Anatomy and examination/diagnosis. *J. Man. Manip. Ther.* **2014**, *22*, 2–12. [CrossRef] [PubMed]
2. Voog, O.; Alstergren, P.; Leibur, E.; Kallikorm, R.; Kopp, S. Impact of temporomandibular joint pain on activities of daily living in patients with rheumatoid arthritis. *Acta Odontol. Scand.* **2003**, *61*, 278–282. [CrossRef] [PubMed]
3. Helenius, L.M.; Tervahartiala, P.; Helenius, I.; Al-Sukhun, J.; Kivisaari, L.; Suuronen, R.; Kautiainen, H.; Hallikainen, D.; Lindqvist, C.; Leirisalo-Repo, M. Clinical radiographic and MRI findings of the temporomandibular joint in patients with different rheumatic diseases. *Int. J. Oral Max. Surg.* **2006**, *35*, 983–989. [CrossRef] [PubMed]
4. Cunha, S.C.d.; Nogueira, R.V.B.; Duarte, Â.P.; Vasconcelos, B.C.d.E.; Almeida, R.d.A.C. Analysis of helkimo and craniomandibular indexes for temporomandibular disorder diagnosis on rheumatoid arthritis patients. *Rev. Bras. Otorrinolaringol.* **2007**, *73*, 19–26. [CrossRef]
5. Covert, L.; Mater, H.V.; Hechler, B.L. Comprehensive management of rheumatic diseases affecting the temporomandibular joint. *Diagnostics* **2021**, *11*, 409. [CrossRef]
6. Tognini, F.; Manfredini, D.; Montagnani, G.; Bosco, M. Is clinical assessment valid for the diagnosis of temporomandibular joint disk displacement? *Minerva Stomatol.* **2004**, *53*, 439–448.
7. Kijowski, R.; Blankenbaker, D.; Stanton, P.; Fine, J.; De Smet, A. Correlation between radiographic findings of osteoarthritis and arthroscopic findings of articular cartilage degeneration within the patellofemoral joint. *Skelet. Radiol.* **2006**, *35*, 895–902. [CrossRef]
8. Westesson, P.-L. Reliability and validity of imaging diagnosis of temporomandibular joint disorder. *Adv. Dent. Res.* **1993**, *7*, 137–151. [CrossRef]
9. Ardic, F.; Gokharman, D.; Atsu, S.; Guner, S.; Yilmaz, M.; Yorgancioglu, R. The comprehensive evaluation of temporomandibular disorders seen in rheumatoid arthritis. *Aust. Dent. J.* **2006**, *51*, 23–28. [CrossRef]
10. Krishnamoorthy, B.; Mamatha, N.; Kumar, V.A. TMJ imaging by CBCT: Current scenario. *Ann. Maxillofac. Surg.* **2013**, *3*, 80. [CrossRef]
11. Shahidi, S.; Salehi, P.; Abedi, P.; Dehbozorgi, M.; Hamedani, S.; Berahman, N. Comparison of the bony changes of TMJ in patients with and without TMD complaints using CBCT. *J. Dent.* **2018**, *19*, 142.

12. Tsiklakis, K.; Syriopoulos, K.; Stamatakis, H. Radiographic examination of the temporomandibular joint using cone beam computed tomography. *Dentomaxillofac. Radiol.* **2004**, *33*, 196–201. [CrossRef] [PubMed]
13. Gynther, G.; Tronje, G. Comparison of arthroscopy and radiography in patients with temporomandibular joint symptoms and generalized arthritis. *Dentomaxillofac. Radiol.* **1998**, *27*, 107–112. [CrossRef] [PubMed]
14. Alexiou, K.; Stamatakis, H.; Tsiklakis, K. Evaluation of the severity of temporomandibular joint osteoarthritic changes related to age using cone beam computed tomography. *Dentomaxillofac. Radiol.* **2009**, *38*, 141–147. [CrossRef]
15. De Carvalho, A. Discriminative power of Larsen's grading system for assessing the course of rheumatoid arthritis. *Acta Radiol. Diagn.* **1981**, *22*, 77–80. [CrossRef]
16. Ahmad, M.; Hollender, L.; Anderson, Q.; Kartha, K.; Ohrbach, R.; Truelove, E.L.; John, M.T.; Schiffman, E.L. Research diagnostic criteria for temporomandibular disorders (RDC/TMD): Development of image analysis criteria and examiner reliability for image analysis. *Oral Surg. Oral Med. Oral Pathol. Oral Radiol. Endodontol.* **2009**, *107*, 844–860. [CrossRef]
17. Landis, J.R.; Koch, G.G. An application of hierarchical kappa-type statistics in the assessment of majority agreement among multiple observers. *Biometrics* **1977**, *33*, 363–374. [CrossRef]
18. Shim, J.S.; Kim, C.; Ryu, J.J.; Choi, S.J. Correlation between TM joint disease and rheumatic diseases detected on bone scintigraphy and clinical factors. *Sci. Rep.* **2020**, *10*, 4547. [CrossRef]
19. Bessa-Nogueira, R.V.; do Egito Vasconcelos, B.C.; Duarte, A.P.; Góes, P.S.; Bezerra, T.P. Targeted assessment of the temporomandibular joint in patients with rheumatoid arthritis. *J. Oral Maxillofac. Surg.* **2008**, *66*, 1804–1811. [CrossRef]
20. Celiker, R.; Gökçe-Kutsal, Y.; Eryilmaz, M. Temporomandibular joint involvement in rheumatoid arthritis: Relationship with disease activity. *Scand. J. Rheumatol.* **1995**, *24*, 22–25. [CrossRef]
21. Youssef Mohamed, M.M.; Dahaba, M.M.; Farid, M.M.; Ali Elsayed, A.M. Radiographic changes in TMJ in relation to serology and disease activity in RA patients. *Dentomaxillofac. Radiol.* **2020**, *49*, 20190186. [CrossRef]
22. Wenneberg, B.; Kopp, S. Clinical findings in the stomatognathic system in ankylosing spondylitis. *Eur. J. Oral Sci.* **1982**, *90*, 373–381. [CrossRef]
23. Dervis, E.; Dervis, E. The prevalence of temporomandibular disorders in patients with psoriasis with or without psoriatic arthritis. *J. Oral Rehabil.* **2005**, *32*, 786–793. [CrossRef]
24. Mayes, M.D.; Lacey, J.V., Jr.; Beebe-Dimmer, J.; Gillespie, B.W.; Cooper, B.; Laing, T.J.; Schottenfeld, D. Prevalence, incidence, survival, and disease characteristics of systemic sclerosis in a large US population. *Arthritis Rheum. Off. J. Am. Coll. Rheumatol.* **2003**, *48*, 2246–2255. [CrossRef]
25. Chebbi, R.; Khalifa, H.B.; Dhidah, M. Temporomandibular joint disorder in systemic sclerosis: A case report. *Pan Afr. Med. J.* **2016**, *25*, 164. [CrossRef]
26. Leung, W.K.; Chu, C.H.; Mok, M.Y.; Yeung, K.S.; Ng, S.K. Periodontal status of adults with systemic sclerosis: Case-control study. *J. Periodontol.* **2011**, *82*, 1140–1145. [CrossRef]
27. Kianmehr, N. Update in Systemic Lupus Erythematosus. *Razi J. Med. Sci.* **2018**, *24*, 63–69.
28. Jonsson, R.; Lindvall, A.M.; Nyberg, G. Temporomandibular joint involvement in systemic lupus erythematosus. *Arthritis Rheum. Off. J. Am. Coll. Rheumatol.* **1983**, *26*, 1506–1510. [CrossRef]
29. Crincoli, V.; Piancino, M.G.; Iannone, F.; Errede, M.; Di Comite, M. Temporomandibular Disorders and Oral Features in Systemic Lupus Erythematosus Patients: An Observational Study of Symptoms and Signs. *Int. J. Med. Sci.* **2020**, *17*, 153–160. [CrossRef]
30. Yilmaz, H.H.; Yildirim, D.; Ugan, Y.; Tunc, S.E.; Yesildag, A.; Orhan, H.; Akdag, C. Clinical and magnetic resonance imaging findings of the temporomandibular joint and masticatory muscles in patients with rheumatoid arthritis. *Rheumatol. Int.* **2012**, *32*, 1171–1178. [CrossRef]
31. Wenneberg, B.; Könönen, M.; Kallenberg, A. Radiographic Changes in the Temporomandibular Joint of Patients with Rheumatoid Arthritis, Psoriatic Arthritis, and Ankylosing Spondylitis. *J. Craniomandib. Disord.* **1990**, *4*, 35–39.
32. Konttinen, Y.; Tuominen, T.; Piirainen, H.; Könönen, M.; Wolf, J.; Hietanen, J.; Malmström, M. Signs and symptoms in the masticatory system in ten patients with mixed connective tissue disease. *Scand. J. Rheumatol.* **1990**, *19*, 363–373. [CrossRef]
33. Melchiorre, D.; Calderazzi, A.; Maddali Bongi, S.; Cristofani, R.; Bazzichi, L.; Eligi, C.; Maresca, M.; Ciompi, M. A comparison of ultrasonography and magnetic resonance imaging in the evaluation of temporomandibular joint involvement in rheumatoid arthritis and psoriatic arthritis. *Rheumatology* **2003**, *42*, 673–676. [CrossRef]
34. Lin, Y.-C.; Hsu, M.-L.; Yang, J.-S.; Liang, T.-H.; Chou, S.-L.; Lin, H.-Y. Temporomandibular joint disorders in patients with rheumatoid arthritis. *J. Chin. Med. Assoc.* **2007**, *70*, 527–534. [CrossRef]
35. Larheim, T.A.; Smith, H.-J.; Aspestrand, F. Rheumatic disease of the temporomandibular joint: MR imaging and tomographic manifestations. *Radiology* **1990**, *175*, 527–531. [CrossRef]
36. Kim, Y.; Kim, K.-H.; Ok, S.-M.; Ahn, Y.-W.; Jeong, S.-H. Conservative Treatment with Occlusal Appliance for Temporomandibular Disorder Patients with Rheumatoid Arthritis. *J. Oral Med. Pain* **2016**, *41*, 169–179. [CrossRef]
37. Wilkes, C.H. Internal Derangements of the Temporomandibular-Joint—Pathological Variations. *Arch. Otolaryngol.* **1989**, *115*, 469–477. [CrossRef]
38. Crow, H.; Parks, E.; Campbell, J.; Stucki, D.; Daggy, J. The utility of panoramic radiography in temporomandibular joint assessment. *Dentomaxillofac. Radiol.* **2005**, *34*, 91–95. [CrossRef]
39. Barghan, S.; Tetradis, S.; Mallya, S. Application of cone beam computed tomography for assessment of the temporomandibular joints. *Aust. Dent. J.* **2012**, *57*, 109–118. [CrossRef]

40. Tasaki, M.M.; Westesson, P.-L. Temporomandibular joint: Diagnostic accuracy with sagittal and coronal MR imaging. *Radiology* **1993**, *186*, 723–729. [CrossRef]

41. Honda, K.; Larheim, T.; Maruhashi, K.; Matsumoto, K.; Iwai, K. Osseous abnormalities of the mandibular condyle: Diagnostic reliability of cone beam computed tomography compared with helical computed tomography based on an autopsy material. *Dentomaxillofac. Radiol.* **2006**, *35*, 152–157. [CrossRef]

42. Librizzi, Z.T.; Tadinada, A.S.; Valiyaparambil, J.V.; Lurie, A.G.; Mallya, S.M. Cone-beam computed tomography to detect erosions of the temporomandibular joint: Effect of field of view and voxel size on diagnostic efficacy and effective dose. *Am. J. Orthod. Dentofac. Orthop.* **2011**, *140*, e25–e30. [CrossRef]

Systematic Review

Psychological Outcomes on Anxiety and Depression after Interventions for Temporomandibular Disorders: A Systematic Review and Meta-Analysis

Lai Ying Luo [1], Johyun Lee [2], Kar Yan Li [3], Yiu Yan Leung [1] and Dion Tik Shun Li [1,*]

1   Oral and Maxillofacial Surgery, Faculty of Dentistry, The University of Hong Kong, Hong Kong SAR, China
2   Faculty of Dentistry, The University of Hong Kong, Hong Kong SAR, China
3   Clinical Research Centre, Faculty of Dentistry, The University of Hong Kong, Hong Kong SAR, China
*   Correspondence: diontsli@hku.hk

Abstract: Many studies have shown mutual interaction between temporomandibular disorders (TMD) and psychological distress. However, evidence on the effectiveness of therapeutic interventions for TMD on psychological outcomes is scarce. This review aimed to summarise the best evidence on the association between interventions for TMD and psychological outcomes regarding symptoms of anxiety and depression. Electronic search was carried out in databases, including Pubmed, Web of Science, Medline, Cochrane Library, and Scopus. All eligible studies were included for narrative synthesis. Eligible randomised controlled trials (RCTs) were included for the meta-analysis. The overall effect size of interventions for TMD was analysed in standardised mean difference (SMD) in levels of anxiety and depression. Ten studies were included in the systematic review. Of these, nine were included in the narrative analysis and four were included in the meta-analysis. All included studies and the result of the narrative analysis showed a statistically significant beneficial effect of interventions for TMD on improving symptoms of anxiety and depression ($p < 0.0001$); however, a statistically significant overall effect was not found in the meta-analyses. Current evidence is in favour of the interventions for TMD in improving symptoms of depression and anxiety. However, the effect is statistically uncertain and warrants future studies to enable the best synthesis of the evidence.

Keywords: temporomandibular joint disorders; depression; anxiety disorders; psychological distress

check for updates

Citation: Luo, L.Y.; Lee, J.; Li, K.Y.; Leung, Y.Y.; Li, D.T.S. Psychological Outcomes on Anxiety and Depression after Interventions for Temporomandibular Disorders: A Systematic Review and Meta-Analysis. *Diagnostics* **2023**, *13*, 653. https://doi.org/10.3390/diagnostics13040653

Academic Editor: Luis Eduardo Almeida

Received: 29 December 2022
Revised: 2 February 2023
Accepted: 5 February 2023
Published: 9 February 2023

1. Introduction

Temporomandibular disorders (TMD) are commonly defined as a group of orofacial disorders involving the masticatory muscles, the temporomandibular joint (TMJ), and adjacent structures with traumatic, neoplastic, and/or musculoskeletal disorders as aetiology [1,2]. Patients often present with a wide and complex range of clinical conditions, including painful conditions, such as myalgia, arthralgia, referred pain, and headache attributed to TMD, and non-painful conditions, such as disc displacement, limited opening, degenerative joint disease, and subluxation [3].

TMD affects 5–15% of adults in general, as reported in different studies, while TMD-related symptoms have been reported to be up to 50% of adults [4]. A recent systematic review and meta-analysis has reported the overall prevalence of TMD diagnosed by the research diagnostic criteria (RDC/TMD) or diagnostic criteria (DC/TMD) to be approximately 31% for adults and elderly [5].

TMD is a common orofacial pain disease, which affects a significant percentage of the population, yet its diagnosis and management remain a challenge. There is a lack of consensus in many aspects because of its multifactorial aetiologies. Although the aetiology of TMD is complex and still not clearly understood, it is generally believed to comprise of biological, psychological, and social factors [6,7]. Therefore, it is important to also consider the psychological symptoms during diagnosis of the disease. The design of DC/TMD and RDC/TMD, the two most widely accepted and standardised assessment tools, has validated the importance of psychological assessment by including psychosocial (Axis II) diagnoses in the dual-axis biopsychosocial diagnostic tool [3].

Depressive and anxiety disorders are the two most common mental disorders, affecting 280 and 301 million people around the world, respectively [8]. Depression is characterised by depressed mood, loss of pleasure or interest in activities, poor concentration, low self-worth, disrupted sleep, change of appetite or weight, and low energy. Patients suffering from depression have a higher risk of committing suicide. Anxiety disorder is characterised by excessive fear, panic attacks, worry in social situations, sleep disturbance, fatigue, sense of tension, nervousness, and restlessness [9,10].

Over the decades, many studies have demonstrated positive correlations between TMD and symptoms of anxiety and depression. In the systematic review of De La Torre Canales et al., a high prevalence of moderate–to–severe depression was observed to range from 21.4 to 60.1% in patients diagnosed with TMD [11]. According to Florjański et al.'s recent literature review, despite the correlation between anxiety and TMD being more controversial when compared to that of depression, the higher prevalence of trait-anxiety (one subtype of anxiety) among patients with TMD than healthy individuals was consistent [9].

The role of a dental surgeon is to detect any symptoms of depression and/or anxiety in patients diagnosed with TMD rather than to diagnose a mental disorder. The most used screening tools, such as the Beck Depression and Anxiety Inventories and the Symptom Checklist-90-Reivsed (SCL-90), are generally questionnaires to reflect patients' self-reported measures.

TMD being a significant and complex health issue, debates are not only over the aetiologies, but also its management. Treatment options range from conservative measures, such as analgesics, occlusal splints, and physiotherapy, to minimally invasive options, such as arthroscopy, arthrocentesis, and intra-articular injection, to open joint surgery. There are also non-standard treatment options, such as Botox injection, acupuncture, and extracorporeal shockwave therapy [4].

Despite the wide variety of options for intervention, there is an increasing consensus on using a multimodal approach in the management of TMD. More studies have supported the concept of the more comprehensive biopsychosocial model of aetiology instead of the more narrowly focused historical biomedical model, especially for providing an integrated and hence successful management of the disease [12]. It is emphasised to manage TMD as a multidimensional chronic illness by a rehabilitation approach that allows integrated assessment between physical and psychological symptoms, and to treat not only the "disorder", but also the "illness" [13]. Therefore, it is essential to evaluate the efficacy of therapeutic intervention for TMD towards not just the primary treatment outcome, but also the secondary psychological outcomes.

In the literature regarding TMD, most studies have investigated the prevalence and aetiology of the disease. There has been increasing evidence of concurrence and mutual interaction between TMD and anxiety and depression [14]. It is reasonable to suggest that a successful intervention for TMD might improve patients' depression and anxiety symptoms. However, we found no reviews that evaluate the influence of treatments of TMD towards the psychological conditions of the patients.

The objective of this systematic review and meta-analysis was to summarise the best evidence on the association between psychological status (i.e., anxiety and depression) and the outcome of therapeutic interventions for TMD.

2. Materials and Methods

This systematic review was conducted following the Preferred Reporting Items for Systematic Reviews and Meta-Analyses (PRISMA) guidelines. The research protocol was registered with the International Prospective Register of Systematic Reviews (PROSPERO), number CRD42022324116 †‡ († The protocol was registered and published during the period of COVID-19 pandemic. Submissions which passed a basic automated check were published automatically after 30 days of waiting time, in order to allow the PROSPERO team to focus on COVID-19 related reviews. Eligibility of this protocol was not checked by the PROSPERO team before this study was commenced. ‡ The registered protocol was amended to also include studies without control groups in order to increase the variety of studies to review).

2.1. Study Selection

Population
Studies reporting adult patients diagnosed with TMD using the RDC/TMD (Axis I and/or Axis II) or its revisions or the new DC/TMD instruments were included. Studies of patients diagnosed with pain disorders other than TMD were excluded.

Intervention
All standard treatment options for TMD identified with the goal to improve the disease by reducing pain and/or improving jaw function were included if they were systematically delivered to the subjects according to a pre-defined algorithm or protocol and were started and completed during the perioperative period of the studies. These included conservative options, including medications (such as analgesics, non-steroidal anti-inflammatory drugs, and muscle relaxants), occlusal appliances of various designs, physiotherapy (such as muscle training and massage), changing of behaviour (soft diet and rest), minimally invasive options (such as arthroscopy, arthrocentesis and intra-articular injections, and open joint surgical options (such as disc repositioning procedures, removal of osteophytes, removal of pathologic tissue, biopsy of the TMJ and alloplastic replacement of the TMJ). Botox injection, acupuncture, extracorporeal shock wave therapy, and laser auriculotherapy, which are currently not considered standard treatment options of TMD, were excluded. Psychological interventions, such as anti-depressants, counselling, stress coping strategies, etc., were not defined as interventions for TMD in this review.

Controls
Studies that have reported comparative groups of subjects receiving no treatments, placebo treatments, or interventions other than the standard treatment options for TMD mentioned above were categorised as studies with control groups. These comparative groups were analysed under the same subgroup in the meta-analysis.

Outcome
Studies included had to report on psychological outcome regarding the severity of anxiety or/and depression. Assessment tools of anxiety included the State-Trait Anxiety Inventory (STAI), Beck Anxiety Inventory (BAI), Hospital Anxiety and Depression Scale (HADS), General Anxiety Disorder-7 (GAD-7) and SCL-90, while those of depression included the Beck Depression Inventory (BDI), Center for Epidemiologic Studies Depression Scale (CES-D), Geriatric Depression Scale (GDS), Patient Health Questionnaire-9 (PHQ-9), HADS and SCL-90.

2.2. Summary of Eligible Criteria

Inclusion criteria
1. Studies conducted among subjects diagnosed with TMD
2. At least one intervention for TMD was delivered
3. Studies reporting outcomes on depression/anxiety after TMD interventions
4. Studies in the English language

Exclusion criteria
1. Studies in animals
2. Studies conducted in children/adolescents aged below 18
3. Studies conducted in patients with other pain disorders, except TMD
4. Studies not using DC/TMD or RDC/TMD for definitive diagnosis
5. Articles with incomplete information
6. Systematic reviews/meta-analyses, meeting/congress reports, and retrospective studies

2.3. Search Strategy

Electronic search was carried out in databases, including Pubmed, Web of Science, Medline, Cochrane Library, and Scopus. The literature search was constructed around search terms for "TMD", "depression", and "anxiety" (Table 1). No restrictions were considered regarding publication year or language. Titles and/or abstracts were reviewed after the elimination of duplicates to exclude seemingly irrelevant articles. Manual search was then performed through the bibliographical references of these articles. These potentially relevant articles were further screened by applying the inclusion and exclusion criteria mentioned above by two independent reviewers. A third independent reviewer (a senior researcher) was consulted on any cases of persisting disagreement. The total search of all databases was performed within March 2022.

Table 1. Databases searched, search terms used, and number of articles found per database.

Database Searched	Search Terms	Articles Retrieved
Pubmed	("Temporomandibular Joint Disorders" or "Temporomandibular joint disorder" or "TMJ Disorders" or "TMJ Disorder" or "Temporomandibular Disorders" or "Temporomandibular Disorder" or "Temporomandibular Joint Diseases" or "Temporomandibular Joint Disease" or "TMJ Diseases" or "TMJ Disease" or "Temporomandibular joint dysfunction syndrome" or "Temporomandibular joint pain" or "Temporomandibular pain" or "TMD" or "Craniomandibular Disorders" or "Craniomandibular Disorder" or "Orofacial Pain" or "Craniofacial pain") AND ("Depression" or "depressive disorders" or "depression symptoms" or "anxiety" or "mood disorders" or "psychological distress")	1285
Web of Science	("Temporomandibular Joint Disorders" or "Temporomandibular joint disorder" or "TMJ Disorders" or "TMJ Disorder" or "Temporomandibular Disorders" or "Temporomandibular Disorder" or "Temporomandibular Joint Diseases" or "Temporomandibular Joint Disease" or "TMJ Diseases" or "TMJ Disease" or "Temporomandibular joint dysfunction syndrome" or "Temporomandibular joint pain" or "Temporomandibular pain" or "TMD" or "Craniomandibular Disorders" or "Craniomandibular Disorder" or "Orofacial Pain" or "Craniofacial pain") AND ("Depression" or "depressive disorders" or "depression symptoms" or "anxiety" or "mood disorders" or "psychological distress")	1387
Medline	(Temporomandibular Joint Disorders or Temporomandibular joint disorder or TMJ Disorders or TMJ Disorder or Temporomandibular Disorders or Temporomandibular Disorder or Temporomandibular Joint Diseases or Temporomandibular Joint Disease or TMJ Diseases or TMJ Disease or Temporomandibular joint dysfunction syndrome or Temporomandibular joint pain or Temporomandibular pain or TMD or Craniomandibular Disorders or Craniomandibular Disorder or Orofacial Pain or Craniofacial pain) and (Depression or depressive disorders or depression symptoms or anxiety or mood disorders or psychological distress)	1027
Cochrane	("Temporomandibular Joint Disorders" or "Temporomandibular joint disorder" or "TMJ Disorders" or "TMJ Disorder" or "Temporomandibular Disorders" or "Temporomandibular Disorder" or "Temporomandibular Joint Diseases" or "Temporomandibular Joint Disease" or "TMJ Diseases" or "TMJ Disease" or "Temporomandibular joint dysfunction syndrome" or "Temporomandibular joint pain" or "Temporomandibular pain" or "TMD" or "Craniomandibular Disorders" or "Craniomandibular Disorder" or "Orofacial Pain" or "Craniofacial pain") AND ("Depression" or "depressive disorders" or "depression symptoms" or "anxiety" or "mood disorders" or "psychological distress")	237
Scopus	("Temporomandibular Joint Disorders" OR "Temporomandibular joint disorder" OR "TMJ Disorders" OR "TMJ Disorder" OR "Temporomandibular Disorders" OR "Temporomandibular Disorder" OR "Temporomandibular Joint Diseases" OR "Temporomandibular Joint Disease" OR "TMJ Diseases" OR "TMJ Disease" OR "Temporomandibular joint dysfunction syndrome" OR "Temporomandibular joint pain" OR "Temporomandibular pain" OR "TMD" OR "Craniomandibular Disorders" OR "Craniomandibular Disorder" OR "Orofacial Pain" OR "Craniofacial pain") AND ("depression" OR "depressive disorders" OR "depression symptoms" OR "anxiety" OR "mood disorders" OR "psychological distress")	1656

2.4. Data Management

The full texts of the articles included were retrieved. Detailed data were extracted from articles independently by two authors according to the data collection form, including information on the author, year of publication, country of publication, study design, size of the population at baseline, characteristics of the population (age at baseline, distribution of experimental, and control groups), duration of follow-up, diagnostic tools of TMD, types of interventions for TMD, outcome measure of TMD intervention, assessment tools of anxiety or/and depression, number of subjects included in the analysis (number of subjects in total, experimental, and control groups), change in treatment outcome of TMD, and severity of anxiety or/and depression before and after interventions.

2.5. Assessment of Risk of Bias and Quality Evaluation

Risks of bias were independently rated by two reviewers based on version 2 of the Cochrane risk-of-bias tool for randomised trials (RoB 2) for randomised controlled studies, based on five domains: bias arising from the randomization process; bias due to deviations from intended interventions; bias due to missing outcome data; bias in measurement of the outcome; and bias in selection of the reported result. A risk-of-bias judgement was reached for each domain, then an overall judgment, by assigning one of the three levels: low risk of bias; some concerns; or high risk of bias [15].

A modified Newcastle–Ottawa Quality Assessment Scale was designed to evaluate the quality of all studies included in this review, with reference to the original assessment scale for cohort studies [16]. A "star system" was employed to judge each study based on three main domains: the selection of the sample, the comparability of the groups, and the ascertainment of the outcome. A maximum of three stars for "Selection", one star for "Comparability", and three stars for "Outcome", which made up a maximum of seven stars that could be scored by each study. This modified questionnaire was designed to provide a quick and direct critical appraisal of the included studies. A study with a total score of 6–7 was categorised as good quality, 3–5 as fair quality, and 0–2 as poor quality. The detailed questionnaire is available in Appendix A.

A third independent reviewer (a senior researcher) was consulted on any discrepancies until consensus was reached.

2.6. Data Analysis

The meta-analyses were performed using the Review Manager (RevMan) 5 software (Version 5.4, The Nordic Cochrane Centre, Copenhagen) when at least two studies reporting specific outcomes were available. A fixed effects model was employed because only a small number of studies (i.e., less than five) were eligible to be included in each analysis [17,18]. All p-values were reported, and $p \leq 0.05$ was considered as being statistically significant.

2.7. Meta-Analyses including Only Studies with Control Groups

The effects of interventions for TMD on depression and anxiety, compared to control interventions, were analysed.

Standardised mean difference (SMD) was used as a summary statistic in the meta-analysis since all studies assessed the same outcome, but with various measurement tools (for example, Costa et al. [19] used HADS, while Alajbeg et al. [20] used PHQ-9 in measuring the degree of depression). A SMD allowed standardization of the results of various studies to a uniform scale for analysis. It is calculated as the difference in mean outcomes between the intervention and control groups, divided by the standard deviation (SD) of the outcome among participants, with 95% confidence intervals (CIs) [15]. When the SDs were unavailable, they were estimated by calculation, using standard errors, CIs, t-values, interquartile deviations, and/or the correlation coefficient [15,21]. The correlation coefficient was obtained from calculation using reported data in Alajbeg et al.'s study [18], which was reported in considerable detail. The mean differences, when not reported, were calculated by subtracting the post-intervention measurement from the baseline measurement.

Measurements taking the closest to the beginning and the end of the interventions were chosen for calculation when more than one baseline and/or post-intervention measurement was reported.

A positive SMD was defined to represent the beneficial effects of interventions for TMD compared to the control intervention for all outcomes (e.g., improvement in the levels of pain, depression, and/or anxiety). A combined SMD was computed in RevMan when there were more than one intervention group (for example, in Melo et al.'s study [22], there were three intervention groups: occlusal splint, manual therapy, and combined therapy) using the mean difference and SD of each group [15]. Improvement was defined as reduction in the levels of pain, depression, and/or anxiety in all statistical analyses in this review.

The overall effect size was evaluated by interpreting the SMDs using the Cohen's categories, where SMD = 0.2 to 0.5 represents a small effect, SMD = 0.5 to 0.8 a moderate effect, and SMD > 0.8 a large effect [23].

The certainty of the evidence of each outcome was evaluated by the Grades of Recommendation, Assessment, Development, and Evaluation (GRADE) approach by two independent reviewers. Five GRADE considerations were used for assessment, including risk of bias, consistency of effect, imprecision, indirectness, and publication bias [15,24].

2.8. Assessment of Heterogeneity

The statistical heterogeneity was assessed by a chi-squared (χ^2) test and inconsistency (I^2) statistics. A rough guide to interpret I^2 was as follows: 0 to 40%: might not be important; 30 to 60%: moderate; 50 to 90%: substantial; and 70 to 100%: considerable. Considering the low power of the χ^2 test when only a few studies were included in an analysis, a p-value of ≤ 0.10 was used to indicate significant heterogeneity.

2.9. Narrative Analysis including All Studies

Narrative syntheses of the mean difference between the outcomes before and after interventions in all studies (including those without control groups) were conducted by obtaining the mean change and standard error (SE) in each intervention group. When the SEs were unavailable, they were estimated using the SDs and the sample size of the groups [15]. The findings were interpreted with caution because any placebo effect or effects due to background inclusion were not excluded in these analyses. Neither judgement of the overall effect size nor the certainty of evidence was derived to eliminate possible misinterpretations.

3. Results

3.1. Literature Search

5592 records were retrieved through the electronic search, and 2408 records were screened after the elimination of duplicates. After the review of titles and/or abstracts, 2386 irrelevant records were excluded because their diagnoses for TMD were not by DC/TMD or RDC/TMD or their variations, and/or there were no interventions for TMD carried out. Out of the 22 full texts reviewed, 12 of them were excluded after being assessed for eligibility because either psychological outcomes were not reported [25–31], or no standard interventions for TMD were delivered [32–36]. No additional records were retrieved after manual search through the reference lists of identified articles. Among the 10 studies (8 RCTs and 2 non-randomised clinical trials) included for qualitative review, 1 RCT [37] was excluded from any quantitative analyses because of insufficient statistical details. A total of 9 studies with 713 patients were included in the narrative analysis. Three RCTs were further excluded from the meta-analysis because either all subjects received interventions for TMD, including the control group (i.e., conservative treatments for TMD) [38], or the assigned interventions were not considered to be standard treatment options [39,40]. Finally, 4 RCTs with 203 patients were eligible and included in the meta-analysis (Figure 1).

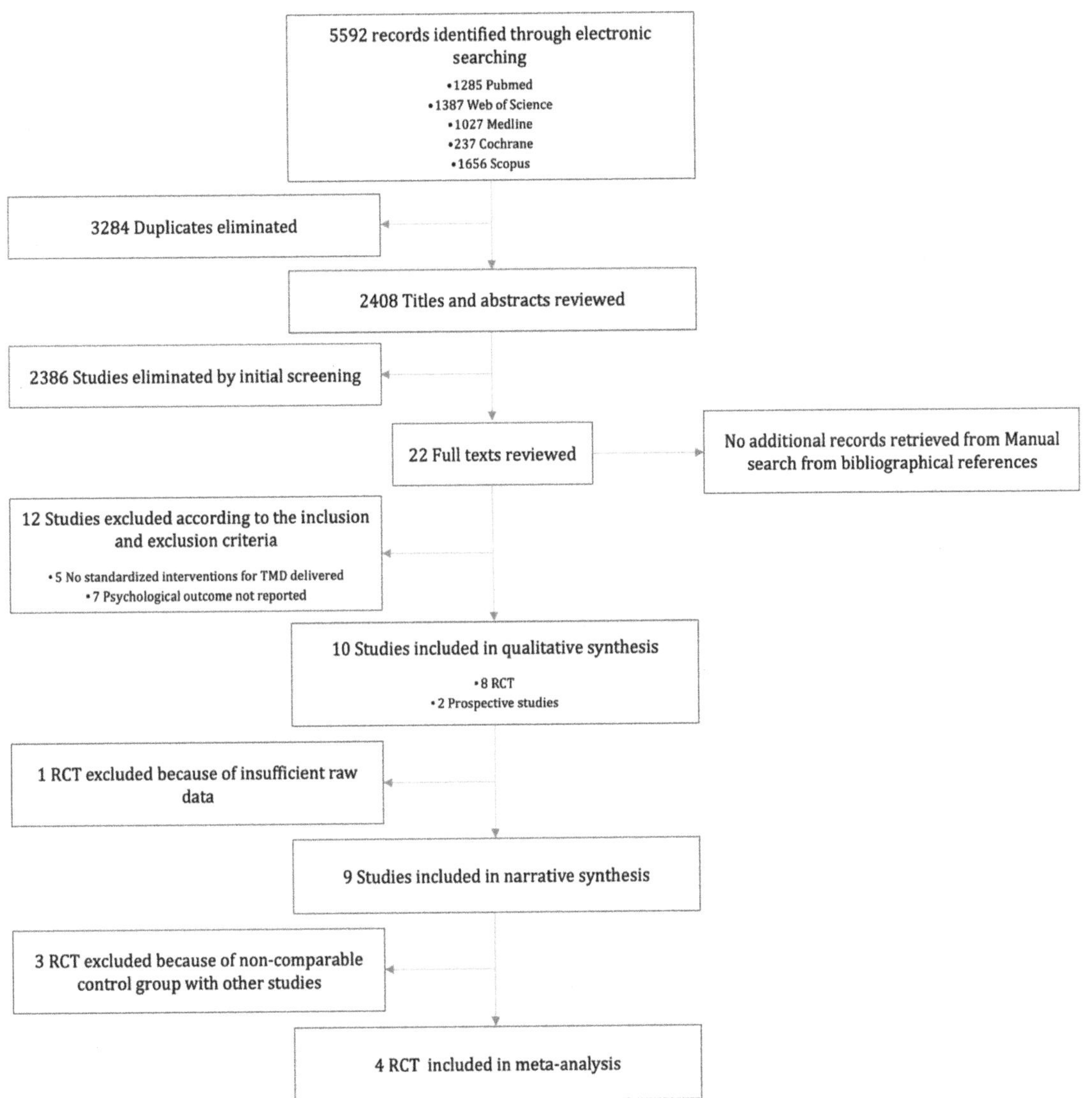

Figure 1. PRISMA flowchart of the result of literature search.

3.2. Study Characteristics

Population characteristics

The summary characteristics of the included studies are presented in Table 2. Of the 10 studies, 5 originated from South America (Brazil) [19,22,37,39,40]; 3 from Europe (Croatia [20], Romania [41], and Portugal [42]); and 2 from North America (USA) [38,43]. The majority of the sample population were made up of patients recruited from dental school clinics [20,22,37,40–43], and the others from private dental clinics [41] and recruitment among local community [19] and primary school teachers [39]. One study only recruited elderly aged 60–79 years [41], while one only recruited female patients [40].

Table 2. Summary of characteristics of included studies.

Study	Country	Characteristics of Subjects - Number of Subjects at Baseline - Age at Baseline - Female to Male Ratio	Intervention Groups - Number of Subjects in Each Group - Description of Interventions - Description of Control (If Any) - Duration of Treatment		Duration of Follow-Up	Diagnostics Tool of TMD	Outcome Measures of Anxiety or/and Depression	Results
			Experimental	Control				
Brandão et al., 2022 [37]	Brazil	- 23 Adults diagnosed with TMD - Recruited from the Center for Health and Functional Studies at the Health Sciences Institute of the Federal University of Bahia - Mean age 35.9 ± 10.5 † - 84.2% Female	- 12 Subjects - Circular massage; at masseter muscle; for 5-min - Pain relief exercise; 30-min session - By researcher; twice weekly; for 4 weeks	- 11 Subjects - Self-care protocol: avoid opening mouth widely, hard food and oral parafunction - Reassess after 30 days	- 1 month	RDC/TMD	Depression: RDC/TMD Axis II	- Improvement in depression demonstrated in the intervention groups with a considerable effect size - Significance not reported
De la Torre Canales et al., 2021 [40]	Brazil	- 20 Female diagnosed with myofascial pain (There were a total of 100 subjects in the paper, evenly distributed in the occlusal appliance group, saline injection group and three groups of botulinum toxin injection in different preparation. However, botulinum injection is defined as non-standardised intervention in this review, only the occlusal appliance group is included in the analysis.) - Recruited from the TMD clinic of Piracicaba Dental School, University of Campinas, São Paulo, Brazil - Mean age 36.8 ± 5.6 - 100% Female	- 20 Subjects - Occlusal appliance - Full coverage; flat; heat-cured acrylic; for upper arch only; canine and anterior guidance occlusal scheme - To wear only during sleep - 6 months	- No control groups	- 6 months	RDC/TMD	Depression: SCL-90R	- Significant improvement in depression demonstrated after treatment

Table 2. *Cont.*

Study	Country	Characteristics of Subjects - Number of Subjects at Baseline - Age at Baseline - Female to Male Ratio	Intervention Groups - Number of Subjects in Each Group - Description of Interventions - Description of Control (If Any) - Duration of Treatment		Duration of Follow-Up	Diagnostics Tool of TMD	Outcome Measures of Anxiety or/and Depression	Results
			Experimental	Control				
Moleirinho-Alves et al., 2021 [42]	Portugal	- 52 Adults diagnosed with myalgia - Recruited from the Egas Moniz University Clinic and the Egas Moniz Dental Clinic - Mean age 25.9 ± 4.5 † - 86.7% Female	- Total 22 Subjects - 15 Subjects in the therapeutic group; physiotherapy consisted of massage of masticatory muscles, isotonic strengthening and coordination exercises by physiotherapist; 30-min session; weekly; for 8 weeks - 17 Subjects in the therapeutic and aerobic exercise group; weekly 30-min physiotherapy and two weekly 30-min aerobic programme; for 8 weeks	- 20 Subjects - Aerobic exercise group - Two weekly cycle ergometer training sessions, for 8 weeks	- 2 months	DC/TMD	Anxiety: GAD-7	- All groups showed small improvement (less than 4 points) in anxiety after treatments Minimum clinically important difference defined as reduction of four points in the total score in GAD-7. - Statistical significance not reported - However, no between-group differences
Alajbeg et al., 2020 [20]	Croatia	- 34 Patients diagnosed with TMD - Recruited from patients seeking treatment for orofacial pain at the School of Dental Medicine, University of Zagreb - Mean age 36.1 ± 11.95 years - 100% Female	- 19 Subjects - Stabilization splint - Full coverage; hard acrylic; for upper arch only; centric relation; 1.5 mm thickness; smooth and flat surface with canine guidance occlusal scheme - To wear it only at night during sleep - 6 months	- 15 Subjects - Placebo splint - A thin transparent foil of 0.5 mm thickness - To wear it only at night during sleep - 6 months	- 3 months - 6 months	DC/TMD	Anxiety: GAD-7 Depression: PHQ-9	- The intervention group presented significantly greater reduction in both anxiety and depression at post-treatment relative to baseline, compared to placebo

Table 2. *Cont.*

Study	Country	Characteristics of Subjects - Number of Subjects at Baseline - Age at Baseline - Female to Male Ratio	Intervention Groups - Number of Subjects in Each Group - Description of Interventions - Description of Control (If Any) - Duration of Treatment		Duration of Follow-Up	Diagnostics Tool of TMD	Outcome Measures of Anxiety or/and Depression	Results
			Experimental	Control				
Melo et al., 2020 [22]	Brazil	- 112 Adults diagnosed with TMD - Recruited by the Department of dentistry, the Federal University of Rio Grande do Norte - Mean age 28 ± 9.34 - 82.1% Female	- Total 84 Subjects - 28 Subjects in the occlusal splint group; full coverage; hard acrylic; for upper arch only; even interocclusal contact in centric relation position; canine guidance occlusal scheme; wear at night and/or daytime; adjustments after 15 days - 28 Subjects in the occlusal splint with counselling group; Reinforcement after 15 days - 28 Subjects in the manual therapy group; heat and cryotherapy, therapeutic exercises; performed by a trained researcher, 40-min session, twice per week, for 4 weeks	- 28 Subjects - Counselling only - Written booklet with dietary guidelines, physical exercises, lifestyle modification - Individualized investigation of possible aetiology and orientated guidelines - Reinforcement after 15 days	- 1 month	RDC/TMD	Anxiety: HADS, BAI	- All groups presented significant improvement in anxiety after 1 month of treatment - However, no between-group differences
Rodrigues et al., 2019 [39]	Brazil	- 20 Adults diagnosed with myofascial pain (There were a total of 40 subjects in the paper, evenly distributed in the laser auriculotherapy (LA) and the occlusal splint (OS) group. However, LA is defined as non-standardised intervention in this review, the 20 subjects in the LA group is excluded from the analysis.) - Recruited from a pool of primary school teachers in the city Campina da Lagoa, Paraná, Brazil - Mean age 43.63 - 100% Female	- 20 Subjects - Occlusal splint - Design of splint not specified - For 8 h overnight - Occlusal adjustment after 2 and 7 days - 8 weeks	- No control groups	- 2 months	RDC/TMD	Depression: SCL-90	- Significant improvement in depression demonstrated at post-treatment

Table 2. *Cont.*

Study	Country	Characteristics of Subjects - Number of Subjects at Baseline - Age at Baseline - Female to Male Ratio	Intervention Groups - Number of Subjects in Each Group - Description of Interventions - Description of Control (If Any) - Duration of Treatment		Duration of Follow-Up	Diagnostics Tool of TMD	Outcome Measures of Anxiety or/and Depression	Results
			Experimental	Control				
Checherita et al., 2018 [41]	Romania	- 107 Elderly (aged 60–79 years) diagnosed with TMD - Recruited from two private dental offices and the Mihail Kogalniceanu Clinical Education Base, of Iasi - Mean age 68.72 ± 8.37 years - 69.1% Female	- 107 Subjects - Anti-inflammatory medication - Ibuprofen 800–1200 mg/day, for 7–14 days - Used the lowest effective dose and shortest treatment duration	- No control groups	- Not specified	RDC/TMD	Depression: GDS	- Depressive manifestation improved at post-treatment, demonstrated as increased proportion of elderly with no depressive symptoms after treatment - Significance not reported
Costa et al., 2015 [19]	Brazil	- 60 Adults diagnosed with masticatory myofascial pain - Recruited from local community through advertisements - Mean age 31.85 ± 7.81 † - 90% Female	- 30 Subjects - Occlusal splint and counselling - Full coverage; hard acrylic; for upper arch only; 2–2.5 mm thickness; smooth and flat surface with anterior guidance occlusal scheme - To wear it only at night during sleep - 5 months	- 30 Subjects - Counselling only - Verbal and written information about TMD aetiology and prognostics, diet modification, lifestyle modification, relaxation of jaw, warm pack, self-massage - 5 months	- 2 months - 5 months	RDC/TMD	Anxiety: HADS, Depression: HADS	- Only the intervention group demonstrated significant reduction in anxiety and depression - However, no between-group differences

Table 2. *Cont.*

Study	Country	Characteristics of Subjects - Number of Subjects at Baseline - Age at Baseline - Female to Male Ratio	Intervention Groups - Number of Subjects in Each Group - Description of Interventions - Description of Control (If Any) - Duration of Treatment		Duration of Follow-Up	Diagnostics Tool of TMD	Outcome Measures of Anxiety or/and Depression	Results
			Experimental	Control				
Turner et al., 2006 [38]	USA	- 158 Adults diagnosed with TMD - Recruited from the University of Washington Orofacial Pain Clinic - Mean age 37.0 ± 11.4 - 86% Female	- Total 158 subjects - All Subjects received conservative interventions for TMD: jaw exercise, warm and/or cold pack, diet modification - Some were prescribed medications and occlusal splints - 79 Subjects in in the pain management training (PMT) group; a brief cognitive-behavioural therapy conducted by a psychologist - 79 subjects in the self-care management (SCM); an education/attention control condition conducted by a bachelor's level educator trained by a psychologist - Biweekly; for 8 weeks	- No control groups	- 2 months - 6 months - 12 months	RDC/TMD	Depression: BD [14]	- Improvement in depression demonstrated in both intervention groups at all post-treatment assessments - Significance not reported - The PMT group presented a significantly higher improvement than the SCM group at 12 months
Rudy et al., 1995 [43]	USA	- 150 Adults diagnosed with TMD - Recruited from an outpatient TMD clinic at the University of Pittsburgh Medical Center - Mean age 32.4 ± 8.4 - 89% Female	- 150 Subjects - Interocclusal appliance and biofeedback/stress management treatment - Full coverage; heat-cured; flat; upper or lower arch; 1–2 mm thickness; even contact in centric relationship; canine guidance occlusal scheme - To wear all times for the first 4 weeks, except during meals and oral hygiene - Biofeedback-assisted relation procedures and stress management treatment conducted by psychologist; 75-min session; weekly; for 6 weeks	- No control groups	- 6 weeks (not reported) - 6 months	RDC/TMD	Depression: BDI	- Significant improvement in depression demonstrated at 6 months follow-up

† Combined mean and SD of age derived from Cochrane's Formula. GAD-7, General Anxiety Disorder-7; PHQ-9, Patient Health Questionnaire-9; GDS, Geriatric Depression Scale; HADS, Hospital Anxiety and Depression Scale; BAI, Beck Anxiety Inventory; SCL-90, Symptoms Checklist-90; BDI, Beck Depression Inventory.

A total of 736 subjects were included in this review. All patients were diagnosed with TMD using RDC/TMD [19,22,37–41,43] or DC/TMD [20,42]. In total, 20% of them were specifically diagnosed with myofascial pain or myalgia [19,39,40,42]. Females made up most of the sample population in all included studies, ranging between 69.1 and 100% (median: 87.9%). The mean age ranged from 25.9 to 68.72 years, with a median age of 36 years. Race was only reported in three studies [20,37,38].

<u>Intervention characteristics</u>

All included studies delivered conservative treatments for TMD, with or without control groups. Most of them used occlusal splint as the major intervention, with adjunct diet and lifestyle modification. A thin (ranged from 1.5–2.5mm), full-coverage upper hard acrylic splint, with even occlusal contact and a canine/anterior guidance occlusal scheme, to be worn only during sleep, was the most common protocol [19,20,40]. One study required patients to wear splints for upper or lower arches at all times, except during meals [43]. One study required patients to wear a splint during the day and/or night [22]. One study did not specify the design of splints [39]. Four studies used massage, a warm pack, and/or cryotherapy at masticatory muscles as interventions [22,37,38,42]. One study used anti-inflammatory medications as the only standardised intervention for TMD [41]. The duration of treatment ranged from 1–6 months (median: 2 months). One study did not specify the duration of treatment [41].

Among the eight RCTs included, only five studies fulfilled the definition of control group in this review. One RCT compared the occlusal splint to the placebo splint with the same wearing schedule [20]. The other four RCTs compared interventions for TMD to other non-standardised treatments, including counselling [19,22], self-care protocol [37], and aerobic exercise [42].

<u>Outcome measures</u>

Two studies [19,20] assessed the severity of both anxiety and depression to evaluate the outcome of interventions, while two [22,42] only assessed the severity of anxiety, and six [37–41,43] only assessed the severity of depression. The level of anxiety was assessed using the General Anxiety Disorder-7 [20,42], Hospital Anxiety and Depression Scale [19,22], and Beck Anxiety Inventory [22]. The level of depression was assessed using Beck's Depression Inventory [38,43], the Symptom Checklist-90-Revised Instrument [37,39,40], the Patient Health Questionnaire-9 [20], and the Geriatric Depression Scale [41].

3.3. Quality of Studies

The quality of the 10 studies included is summarised and presented in Table 3. Three studies [20,22,42] were judged as "good" quality, six studies [19,37–40,43] were judged as "fair quality", and one study [41] was judged as "poor quality". Most of the studies that were judged as "poor" or "fair quality" were due to the lack of representativeness of the sample, small sample size, inadequate follow-up period, or lack of description to data lost.

Table 3. Quality assessment of studies using the modified Newcastle–Ottawa Scale.

Study	Selection	Comparability	Outcome	Total Score
Alajbeg et al., 2020 [20]	***	*	***	7
Melo et al., 2020 [22]	***	*	**	6
Moleirinho-Alves et al., 2021 [42]	***	*	**	6
Costa et al., 2015 [19]	***	*	*	5
De la Torre Canales et al., 2021 [40]	**	*	**	5
Rudy et al., 1995 [43]	**		***	5
Turner et al., 2006 [38]	**	*	**	5
Brandão et al., 2022 [37]	*	*	**	4
Rodrigues et al., 2019 [39]	**	*	*	4
Checherita et al., 2018 [41]			**	2

Total score of 6–7: good quality; 3–5: fair quality; and 0–2: poor quality. *, **, *** Represents the score awarded in each section.

3.4. Risk of Bias in Studies Included in the Meta-Analysis

Only one RCT [20] was judged to have low risk of bias, one RCT [22] was judged to have some concern of bias, while the other two RCTs [19,42] were judged to have high risk of bias. The summary and description of the risk of bias assessment is presented in Figure 2. All studies had a low risk of bias in the measurement of the outcome, as they used common and standardised screening tools for the assessment of anxiety and/or depression with adequate description. The risk of bias in the randomization process was somewhat high because one RCT [22] did not report on adequate allocation concealment, and one RCT [42] did not allocate participants in a randomised manner, but according to participants' preferences. The risk of bias in missing outcome data was high in one RCT [19] because of a high dropout rate of 32%, in which the number of dropped-out participants doubled in the control group compared to the test group, which was likely to induce bias in the result. The risk of bias in the selection of the reported result was generally of some concern or high because the numerical results reported in most of the studies were likely to be selected, such as the mean difference between the test and control groups were not always reported.

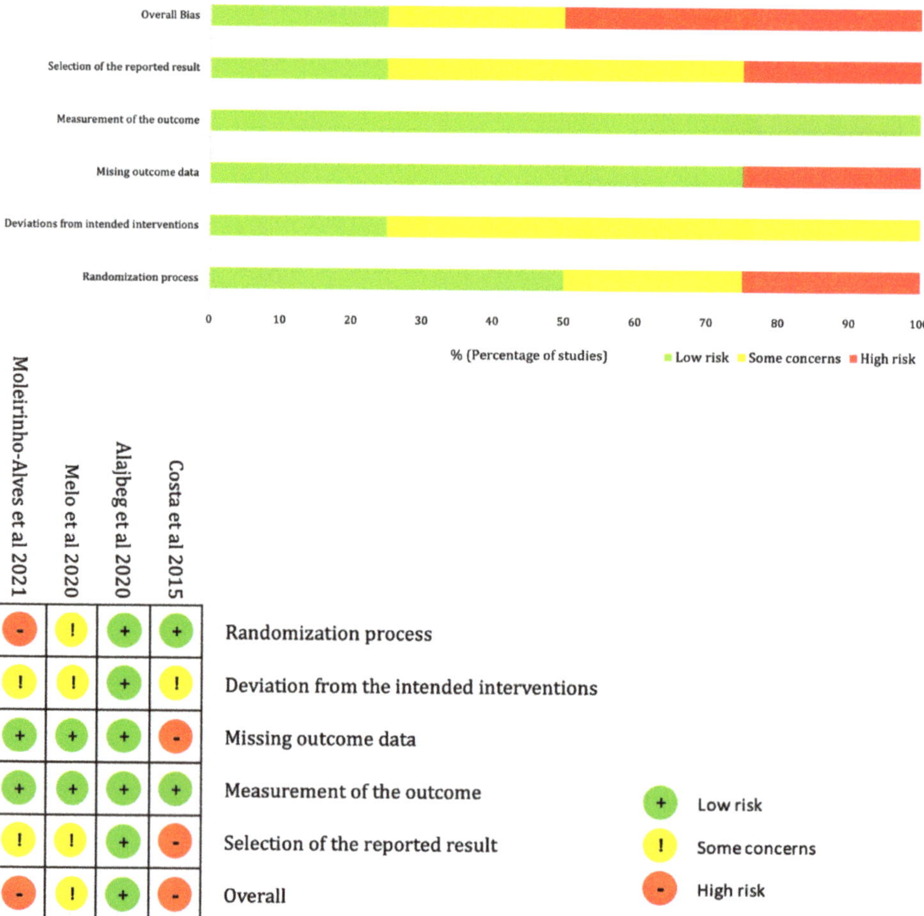

Figure 2. Summary and description of risk of bias assessment of the studies included in the meta-analysis using the Cochrane risk-of-bias tool for randomised trials (RoB 2) [19,20,22,42].

3.5. Narrative Analysis

Nine out of ten studies provided sufficient data regarding anxiety and/or depression to evaluate the overall effects of interventions over time, without controlling for the placebo effect for narrative analyses. Therefore, the results shall be interpreted with caution.

Anxiety

Four studies provided sufficient data regarding anxiety for the narrative analysis [19,20,22,42]. The combined data of the 129 participants who received interventions for TMD showed a statistically significant improvement in the symptoms of anxiety (SMD = 2.15; 95% CL 1.66 to 2.65; $p < 0.00001$). Very low and statistically insignificant heterogeneity was observed between studies (Heterogeneity: $I^2 = 0\%$; $\chi^2 = 2.94$; $p = 0.40$) (Figure 3).

Study or Subgroup	Std. Mean Difference	SE	Weight	Std. Mean Difference IV, Fixed, 95% CI
Melo et al, 2020	1.74	0.35	52.2%	1.74 [1.05, 2.43]
Costa et al, 2015	2.5	0.67	14.2%	2.50 [1.19, 3.81]
Alajbeg et al, 2020	2.63	0.96	6.9%	2.63 [0.75, 4.51]
Moleirinho–Alves et al, 2021	2.65	0.49	26.6%	2.65 [1.69, 3.61]
Total (95% CI)			100.0%	2.15 [1.66, 2.65]

Heterogeneity: Chi² = 2.94, df = 3 (P = 0.40); I² = 0%
Test for overall effect: Z = 8.51 (P < 0.00001)

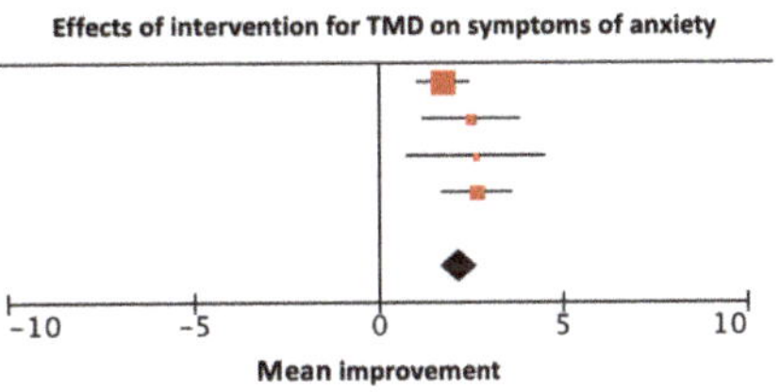

Figure 3. Forest plot of effects of intervention for TMD on symptoms of anxiety. Box size reflects study size. The diamond at the bottom reflects the overall pooled effect with a 95% confident interval. There was an overall significant mean improvement in symptoms of anxiety after interventions for TMD [19,20,22,42].

Depression

Seven studies provided sufficient data regarding depression for the narrative analysis [19,20,38–41,43]. A random-effects model was employed to incorporate heterogeneity because a considerable number of studies were included [15]. The combined data of the 451 participants who received interventions for TMD showed a statistically significant improvement in symptoms of depression (SMD = 1.76; 95% CL 0.94 to 2.59; $p < 0.0001$). Strong and statistically significant heterogeneity was noted between studies (Heterogeneity: $Tau^2 = 0.97$; $I^2 = 88\%$; $\chi^2 = 51.84$; $p < 0.00001$). Subgroup analyses regarding different interventions for TMD were performed.

Statistically significant improvement was observed in all three kinds of interventions. The test for subgroup differences suggested that a statically significant subgroup effect was evidenced (<0.0001). The treatment effect was greater for massage and warm pack or cryotherapy, followed by occlusal splint and analgesics (Massage and warm pack or cryotherapy: SMD = 3.47; 95% CL 2.12 to 4.82; $p < 0.00001$; Occlusal splint: SMD = 1.81; 95% CL 0.64 to 2.98; $p = 0.002$; and Analgesics: SMD = 0.53; 95% CL 0.08 to 0.98; $p = 0.02$). However, there was unexplained heterogeneity between trials within the subgroup of occlusal splint ($Tau^2 = 1.46$; $I^2 = 89\%$; $\chi^2 = 35.17$; $p < 0.0001$), which required further investigation (Figure 4).

Sensitivity analysis

In order to investigate factors contributing to the heterogeneity across studies, sensitivity analyses were performed by repeating the analyses according to the assessment tools used for depression. Mean differences for each tool were individually analysed. Statistically significant improvements in depression remained (BDI: MD = 3.28; 95% Cl 2.53 to 4; $p < 0.00001$; SCL-90R: MD = 0.68; 95% Cl 0.40 to 0.96; $p < 0.00001$), while no heterogeneity was observed within studies using the same assessment tool (BDI: $I^2 = 0\%$; $\chi^2 = 0.11$; $p < 0.74$; SCL-90R: $I^2 = 0\%$; $\chi^2 = 0.43$; $p < 0.51$) (Table 4).

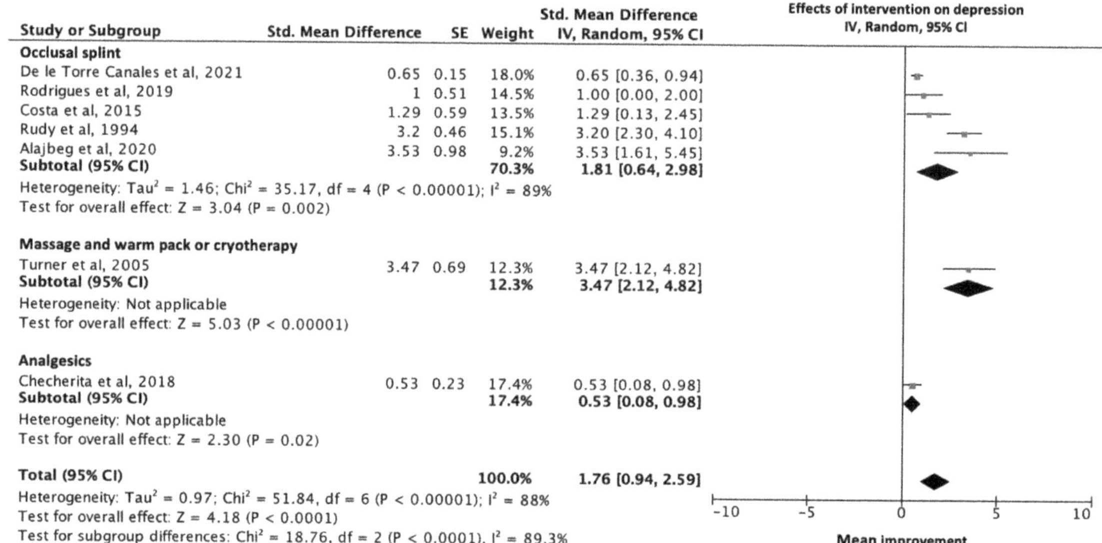

Figure 4. Forest plot of subgroup analysis of effects of intervention for TMD on symptoms of depression according to different interventions. Box size reflects study size. The diamond at the bottom reflects the overall pooled effect with a 95% confident interval. There was an overall significant mean improvement in symptoms of depression after interventions for TMD, as well as significant subgroup differences between different interventions [19,20,38–41,43].

Table 4. Summary of the narrative analysis of effects of interventions for TMD on severity of anxiety and depression and sensitivity analyses.

	Number of Studies	Included Studies	Number of Participants (Intervention)	SMD (95% CI)	*p* Value	Heterogeneity I2; χ2; P
Intervention effects						
Anxiety	4	[19,20,22,42]	139	2.15 (1.66–2.65)	<0.00001	0%; 2.94; 0.40
Depression	7	[19,20,38–41,43]	451	1.76 (0.94–2.59)	<0.0001	88%; 51.84; <0.00001
Sensitivity analysis (Depression)						
Assessed by BDI	2	[38,43]	270	3.28 (2.53, 4.03) *	<0.00001	0%; 0.11; 0.74
Assessed by SCL-90R	2	[39,40]	31	0.68 (0.40, 0.96) *	<0.00001	0%; 0.43; 0.51

* Mean differences instead of SMD were estimated because the same assessment tools were used in the studies included. SMD, Standard mean difference; BDI, Beck's Depression Index; SCL-90R, Screening Checklist—90 Revised.

3.6. Meta-Analysis

Four RCTs out of ten studies provided sufficient data regarding anxiety and/or depression for the meta-analysis to evaluate the overall effects of intervention over time, with the control of placebo effects. The summary of the results of the overall effects of intervention on anxiety and depression compared with the control group and the sensitivity analysis are presented in Table 5.

Table 5. Summary of effects of interventions for TMD on severity of anxiety and depression and sensitivity analysis.

	Number of Studies	Included Studies	Number of Participants (Intervention)	Number of Participants (Control)	SMD (95% CI)	p Value	Heterogeneity I2; χ2; P
Intervention effects							
Anxiety	4	[19,20,22,42]	139	64	0.29 (0.02–0.6)	0.06	0%; 1.80; 0.62
Depression	2	[19,20]	43	32	0.40 (−0.06–0.87)	0.09	0%; 0.22; 0.64
Sensitivity analysis (Anxiety)							
Higher-quality studies	2	[20,22]	90	34	0.11 (−0.3–0.51)	0.06	0%; 0.00; 0.99
Occlusal splints	3	[19,20,22]	109	49	0.22 (−0.13–0.57)	0.22	0%; 1.06; 0.59
Assessed by GAD−7	2	[20,42]	54	32	1.07 (−0.37, 2.52) *	0.15	0%; 0.61; 0.44
Assessed by HADS	2	[19,22]	85	32	0.87 (−0.61, 2.36) *	0.25	23%; 1.29; 0.26

* Mean differences instead of SMD were estimated because the same assessment tools were used in the studies included. SMD, Standard mean difference; GAD-7, General Anxiety Disorder–7; HADS, Hospital Anxiety and Depression Scale.

Anxiety

All four RCTs reported data regarding symptoms of anxiety, as evaluated using the GAD-7 questionnaire [20,42] or HADS [19,22]. The level of anxiety was assessed by two screening tools, HADS and BAI, in Melo's RCT [22]. Data evaluated using HADS was extracted for this meta-analysis to minimise the heterogeneity between different screening tools. Analysis of these 4 studies (139 participants in the intervention arm and 64 participants in the control arm) showed no significant difference between the 2 groups (SMD = 0.29; 95% CL −0.02 to 0.60; p = 0.06) (Figure 5).

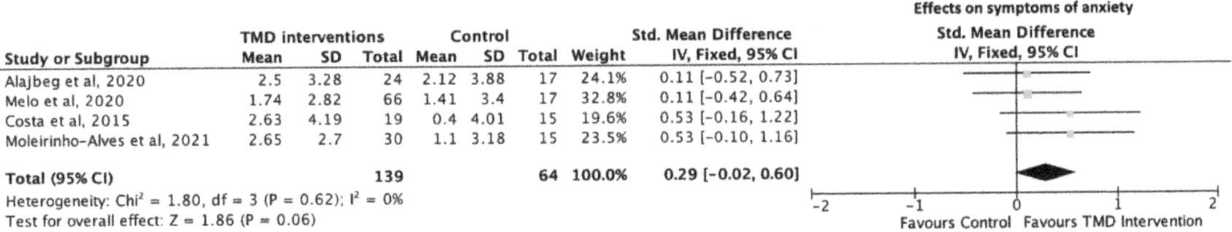

Figure 5. Forest plot of effects of intervention for TMD on symptoms of anxiety after controlling for placebo effect. Box size reflects study size. The diamond at the bottom reflects the overall pooled effect with a 95% confident interval. Positive SMD reflects effect on improving symptoms of anxiety favouring interventions for TMD over control. No statistically significant differences were observed between the two groups [19,20,22,42].

Depression

Only two RCTs reported sufficient data regarding symptoms of depression, as evaluated using the Patient Health Questionnaire-9 [20] or the HADS [19]. Analysis of these 2 studies (43 participants in the intervention arm and 32 participants in the control arm), showed no significant differences between the 2 groups (SMD = 0.40; 95% CL −0.06 to 0.87; p = 0.09) (Figure 6).

Sensitivity analysis

A sensitivity analysis was performed by repeating the meta-analysis regarding the effect of interventions on symptoms of anxiety after removing two studies [19,42] with a high risk of bias. The difference between the intervention and control groups remained insignificant (SMD = 0.11; 95% CL −0.3 to 0.51; p = 0.06). Sensitivity analyses were performed by repeating the meta-analyses, according to the assessment tools used for anxiety. The mean differences for each tool were individually analysed. However, there were still no significant differences between the intervention and control groups observed. Since all the studies delivered occlusal splints in their intervention arms, except Moleirinho-

Alves et al. [42], which used massage and warm pack or cryotherapy as intervention, the analysis was repeated after removing its influence. Similarly, no significant differences between the intervention and control groups were observed (Table 5).

Quality of evidence

The level of certainty of the evidence was judged in the GRADE approach. Despite the low heterogeneity between studies in the analyses regarding both anxiety and depression, there were considerable risks of bias due to the generally small sample size in all studies, lack of blinding in both participants and clinicians in most studies, and high attrition rate in some studies. Therefore, it was deemed appropriate to downgrade the certainty of the evidence by two levels, from high to low, due to the imprecision of the results and the study limitations.

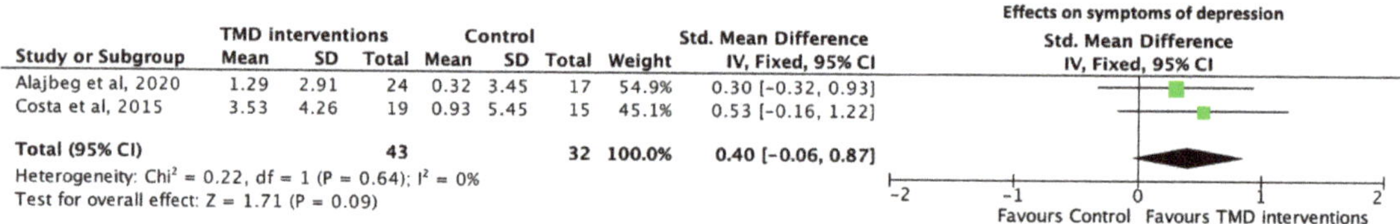

Figure 6. Forest plot of effects of interventions for TMD on symptoms of depression after controlling for placebo effect. Box size reflects study size. The diamond at the bottom reflects the overall pooled effect with a 95% confident interval. Positive SMD reflects effect on improving symptoms of depression favouring interventions for TMD over control. No statistically significant differences were observed between the two groups [19,20].

4. Discussion

4.1. Summary of the Findings

This systematic review and meta-analysis explored the best available evidence on the effectiveness of interventions for TMD on psychological outcome regarding symptoms of anxiety and depression in patients diagnosed with TMD. A total of 10 studies fulfilled the inclusion criteria and underwent qualitative analysis, while 9 studies provided sufficient data for the narrative analysis, and 4 RCTs for the meta-analysis. The results in all the studies generally suggested significant improvement in anxiety and depression after interventions for TMD, which is further demonstrated in our narrative analysis by an overall statistically significant reduction in the level of anxiety and depression. An obvious tendency of overall effects on improving symptoms in both depression and anxiety favouring interventions for TMD over control was observed in the meta-analyses; however, the effectiveness was found not statistically significant regarding a 95% confident interval. Furthermore, the subgroup analysis for the treatment effect on the improvement in depression regarding different interventions showed statically significant group differences, which in turn suggested that different interventions significantly modified the effect on the improvement in symptoms of depression. Heterogeneity was observed within subgroups, which suggested possible background factors that contributed to the varied results. In the sensitivity analysis, no heterogeneity was observed within studies using the same psychological assessment tools, suggesting that the use of various psychological assessment tools might be the reason for the heterogeneity.

4.2. Role of Interventions for TMD in Improving Anxiety and Depression

The statistically significant effect observed in the narrative analysis suggests a beneficial effect of interventions for TMD on reducing levels of depression and anxiety, regardless of the types of interventions given. The mechanism of this beneficial effect was suggested to be associated with the relationship between pain and TMD. Previous studies have indicated the mutual interaction between pain and psychological distresses [14,44] Successful therapeutic treatments in patients with TMD are suggested to have a positive effect in improving symptoms of anxiety and depression by pain management strategies [45].

4.3. Statistical Significance Not Found in Treatment Effect When Compared to Control Group

The overall treatment effect tended to favour interventions for TMD over the control in improving symptoms of depression and anxiety. However, it was not found to be statistically significant. This finding was likely because only a limited number of eligible studies were included in this meta-analysis [46]. Furthermore, most studies [19,22] provided treatments such as counselling to patients in the control group; only one study [20] used placebo splint in the control setup. These non-standardised interventions have likely resulted in a positive effect on the psychological outcomes, which have in turn weakened the effects of the standardised interventions shown in the statistics.

4.4. Implication for Clinical Practice

This review suggested a supportive role of interventions for TMD in improving anxiety and depression. It is demonstrated in the Turner et al. [38] and Costa et al. [19] studies that the combination of treatments for TMD and psychological interventions, such as cognitive-behavioural therapy and counselling, resulted in the best outcome. They believed the involvement of a psychological approach allowed relaxation and better pain-coping strategies which worked hand-in-hand with the standardised interventions in the management of TMD. Previous studies also supported the implication of psychosocial interventions for chronic orofacial pain [47]. On the other hand, it is also important for psychologists to be aware of any signs of TMD in their patients. A timely referral to oral surgeons might help in the management of psychological distress of their patients. A multidisciplinary approach is suggested to best manage this multifactorial illness.

4.5. Implication for Future Research

Future RCTs should ensure the high quality of the methodology and reporting, including larger sample sizes, allocation concealment, control groups with no treatments or placebo treatments, and intention-to-treat analyses. Meta-analyses could be repeated when there are more eligible studies available to improve generalization and obtain an accurate overall treatment effect. Future RCTs could be conducted to compare the effectiveness between standardised interventions for TMD; psychological interventions; and combinations of both and no treatments, on both pain control and psychological outcomes. This requires contributions of expertise from both oral surgery and psychology.

4.6. Strengths and Limitations of This Review

There were several limitations in this review. First, only a small number of studies could be included in this meta-analysis. The pooled sample size was relatively small to identify significant relationships within the dataset.

Secondly, high heterogeneity existed in the various assessment scales of anxiety and depression applied in different studies. Multiple cut-off points were used among studies that used the same assessment tools. The duration of intervention varied, and measurements of outcome parameters were obtained at different time-points across studies. These have made direct comparison of the study outcomes difficult. The summary statistics required for meta-analysis were unavailable in most studies, and much statistical estimation was performed, which might induce inaccuracy in the analysis.

Furthermore, the low methodological quality of the available RCTs might also include bias. Since all the assessment tools of anxiety and depression relied on questionnaires completed by patients, blinding of outcome measurements became impossible. Some studies did not conduct intention-to-treat, but rather per-protocol analyses when there were missing data.

In addition, the studies that fulfilled the inclusion criteria, and thus were included in this systematic review, consisted only of a limited array of the currently available treatment options, such as occlusal splint and anti-inflammatory medications. Studies pertaining to other common interventions for TMD, such as intra-articular injection and arthrocentesis, which also fulfil the inclusion criteria of this systematic review, were not found. It is, therefore, not possible to relate the findings of the current systematic review and meta-analysis to those other common interventions for TMD.

Lastly, the patients included in the studies were mostly psychological healthy individuals with symptoms, but not diagnosed with anxiety and depression. The difference before and after interventions might, therefore, be too small to be reflected in the statistics.

Nevertheless, to the best of our knowledge, this is the first systematic review and meta-analysis to evaluate the effectiveness of interventions for TMD in reducing psychological distress. A comprehensive search of available literature was conducted, with an established review methodology applied, to minimise possible bias. Although only a handful of studies could be included in the meta-analysis, we attempted to summarise the best available evidence and identify the current research gap in this topic. This systematic review and meta-analysis serves as an exploratory review, providing a plausible estimate that could be tested in the future in subsequent reviews of the role of interventions for TMD in correcting psychological stress.

5. Conclusions

This systematic review and meta-analysis have suggested the interventions for TMD may be beneficial in improving symptoms of depression and anxiety, based on the current available evidence. However, the effect is statistically uncertain and warrants future studies to enable the best synthesis of the evidence. Multidisciplinary management, with the input of both the surgeons and the psychologists, is recommended in treating patients presented with TMD and symptoms of psychological distress.

Author Contributions: Conceptualization, L.Y.L., Y.Y.L. and D.T.S.L.; methodology, L.Y.L. and D.T.S.L.; formal analysis, L.Y.L., J.L., K.Y.L. and D.T.S.L.; writing—original draft preparation, L.Y.L.; writing—review and editing, L.Y.L., Y.Y.L. and D.T.S.L.; supervision, Y.Y.L. and D.T.S.L. All authors have read and agreed to the published version of the manuscript.

Funding: This research received no external funding.

Institutional Review Board Statement: Not applicable.

Informed Consent Statement: Not applicable.

Data Availability Statement: Data is available in the manuscript.

Conflicts of Interest: The authors declare no conflict of interest.

Appendix A

Modified Newcastle–Ottawa Quality Assessment Scale
For clinical trials

Note: A study can be awarded a maximum of one star for each numbered item. A maximum of seven stars is available in total, including three stars in "Selection", one star in "Comparability", and three stars in "Outcome".

Selection

1. Representativeness of the sample.

    (a) Truly representative of the average in patients with TMD (Random sampling/population based sampling)*.

    (b) Somewhat representative of the average in patients with TMD (non-random sampling)*.

    (c) Selected group of patients, e.g., volunteer, students, hospital staffs, restricted by gender, etc.

    (d) No description of the sampling strategy.

2. Sample size.

    (a) Justified and satisfactory, that is, to have a clear description of algorithm, the required sample size is derived*.

(b) Not justified.

3. Ascertainment of intervention.

   (a) Protocol of intervention is clearly described. Attempts are made to ensure the intervention is accurately carried out*.

   (b) Protocol of intervention is somewhat described, but no confirmation of whether the intervention is accurately carried out.

   (c) No description of intervention or no attempt to ascertain information on how accurate the intervention is carried out.

Comparability

(a) Presence of a control or comparable group, e.g., placebo, no treatment, or other non-standardised interventions for TMD*.

Outcome

1. Assessment of outcome

   (a) Independent blind assessment*

   (b) Record linkage*

   (c) Self-report*

   (d) No description

2. Was follow-up long enough for outcomes to occur

   (a) Yes (length of intervention and follow-up was at least 6 months)*

   (b) No

3. Adequacy of follow-up

   (a) Complete follow up—all subjects accounted for*

   (b) Number of subjects lost to follow-up is more and unlikely to introduce bias (<20%), or description of those lost is clearly reported*

   (c) >20% of subjects lost to follow-up and no description of those lost

References

1. Cascone, P.; Gennaro, P.; Gabriele, G.; Chisci, G.; Mitro, V.; De Caris, F.; Iannetti, G. Temporomandibular synovial chondromatosis with numerous nodules. *J. Craniofac. Surg.* **2014**, *25*, 1114–1115. [CrossRef] [PubMed]
2. Gauer, R.L.; Semidey, M.J. Diagnosis and treatment of temporomandibular disorders. *Am. Fam. Physician* **2015**, *91*, 378–386.
3. Schiffman, E.; Ohrbach, R.; Truelove, E.; Look, J.; Anderson, G.; Goulet, J.P.; List, T.; Svensson, P.; Gonzalez, Y.; Lobbezoo, F.; et al. Diagnostic Criteria for Temporomandibular Disorders (DC/TMD) for Clinical and Research Applications: Recommendations of the International RDC/TMD Consortium Network* and Orofacial Pain Special Interest Group. *J. Oral Facial Pain Headache* **2014**, *28*, 6–27. [CrossRef]
4. Li, D.T.S.; Leung, Y.Y. Temporomandibular Disorders: Current Concepts and Controversies in Diagnosis and Management. *Diagnostics* **2021**, *11*, 459. [CrossRef] [PubMed]
5. Valesan, L.F.; Da-Cas, C.D.; Réus, J.C.; Denardin, A.C.S.; Garanhani, R.R.; Bonotto, D.; Januzzi, E.; de Souza, B.D.M. Prevalence of temporomandibular joint disorders: A systematic review and meta-analysis. *Clin. Oral Investig.* **2021**, *25*, 441–453. [CrossRef] [PubMed]
6. List, T.; Jensen, R.H. Temporomandibular disorders: Old ideas and new concepts. *Cephalalgia* **2017**, *37*, 692–704. [CrossRef] [PubMed]
7. Zhang, Z.K.; Ma, X.C.; Gao, S.; Gu, Z.Y.; Fu, K.Y. Studies on contributing factors in temporomandibular disorders. *Chin. J. Dent. Res.* **1999**, *2*, 7–20.
8. Global Health Data Exchange. Institute of Health Metrics and Evaluation 2019. Available online: https://vizhub.healthdata.org/gbd-results/ (accessed on 8 June 2022).
9. Florjański, W.; Orzeszek, S. Role of mental state in temporomandibular disorders: A review of the literature. *Dent. Med. Probl.* **2021**, *58*, 127–133. [CrossRef]
10. World Health Organization. Mental Disorders. 2022. Available online: https://www.who.int/news-room/fact-sheets/detail/mental-disorders (accessed on 12 June 2022).
11. De La Torre Canales, G.; Câmara-Souza, M.B.; Muñoz Lora, V.R.M.; Guarda-Nardini, L.; Conti, P.C.R.; Rodrigues Garcia, R.M.; Del Bel Cury, A.A.; Manfredini, D. Prevalence of psychosocial impairment in temporomandibular disorder patients: A systematic review. *J. Oral Rehabil.* **2018**, *45*, 881–889. [CrossRef]

12. Suvinen, T.I.; Reade, P.C.; Kemppainen, P.; Könönen, M.; Dworkin, S.F. Review of aetiological concepts of temporomandibular pain disorders: Towards a biopsychosocial model for integration of physical disorder factors with psychological and psychosocial illness impact factors. *Eur. J. Pain* **2005**, *9*, 613–633. [CrossRef]

13. Engel, G.L. The need for a new medical model: A challenge for biomedicine. *Science* **1977**, *196*, 129–136. [CrossRef] [PubMed]

14. Berger, M.; Oleszek-Listopad, J.; Marczak, M.; Szymanska, J. Psychological aspects of temporomandibular disorders–literature review. *Curr. Issues Pharm. Med. Sci.* **2015**, *28*, 55–59. [CrossRef]

15. Higgins, J.P.; Thomas, J.; Chandler, J.; Cumpston, M.; Li, T.; Page, M.J.; Welch, V.A. *Cochrane Handbook for Systematic Reviews of Interventions*; John Wiley & Sons: Hoboken, NJ, USA, 2019.

16. Wells, G.A.; Shea, B.; O'Connell, D.; Peterson, J.; Welch, V.; Losos, M.; Tugwell, P. The Newcastle-Ottawa Scale (NOS) for Assessing the Quality of Nonrandomised Studies in Meta-Analyses. 2000. Available online: https://www.ohri.ca/programs/clinical_epidemiology/oxford.asp (accessed on 8 June 2022).

17. Murad, M.; Montori, V.; Ioannidis, J.; Prasad, K.; Cook, D.; Guyatt, G. *Fixed-Effects and Random-Effects Models. Users' Guide to the Medical Literature. A Manual for Evidence-Based Clinical Practice*, 3rd ed.; McGraw-Hill: New York, NY, USA, 2015.

18. Tufanaru, C.; Munn, Z.; Stephenson, M.; Aromataris, E. Fixed or random effects meta-analysis? Common methodological issues in systematic reviews of effectiveness. *JBI Evid. Implement.* **2015**, *13*, 196–207. [CrossRef] [PubMed]

19. Costa, Y.M.; Porporatti, A.L.; Stuginski-Barbosa, J.; Bonjardim, L.R.; Conti, P.C. Additional effect of occlusal splints on the improvement of psychological aspects in temporomandibular disorder subjects: A randomized controlled trial. *Arch. Oral Biol.* **2015**, *60*, 738–744. [CrossRef] [PubMed]

20. Alajbeg, I.Z.; Vrbanović, E.; Lapić, I.; Alajbeg, I.; Vuletić, L. Effect of occlusal splint on oxidative stress markers and psychological aspects of chronic temporomandibular pain: A randomized controlled trial. *Sci. Rep.* **2020**, *10*, 10981. [CrossRef] [PubMed]

21. Wan, X.; Wang, W.; Liu, J.; Tong, T. Estimating the sample mean and standard deviation from the sample size, median, range and/or interquartile range. *BMC Med. Res. Methodol.* **2014**, *14*, 135. [CrossRef] [PubMed]

22. Melo, R.A.; de Resende, C.; Rêgo, C.R.F.; Bispo, A.S.L.; Barbosa, G.A.S.; de Almeida, E.O. Conservative therapies to treat pain and anxiety associated with temporomandibular disorders: A randomized clinical trial. *Int. Dent. J.* **2020**, *70*, 245–253. [CrossRef] [PubMed]

23. Cohen, J. *Statistical Power Analysis forthe Behavioural Sciences*; Baskı: Hillsdale, NJ, USA, 1988.

24. GRADE Handbook. Available online: https://gdt.gradepro.org/app/handbook/handbook.html (accessed on 12 June 2022).

25. Badel, T.; Keros, J.; Marotti, M.; Kern, J.; Kocijan, S. Therapy of displaced disk of the temporomandibular joint in relation to anxiety. *Period. Biol.* **2008**, *110*, 101–105.

26. Bouloux, G.F.; Zerweck, A.G.; Celano, M.; Dai, T.; Easley, K.A. Can Preoperative Psychological Assessment Predict Outcomes After Temporomandibular Joint Arthroscopy? *J. Oral Maxillofac. Surg.* **2015**, *73*, 2094–2102. [CrossRef]

27. Huttunen, J.; Qvintus, V.; Suominen, A.L.; Sipilä, K. Role of psychosocial factors on treatment outcome of temporomandibular disorders. *Acta Odontol. Scand.* **2019**, *77*, 119–125. [CrossRef]

28. Krogstad, B.S.; Jokstad, A.; Dahl, B.L.; Soboleva, U. Somatic complaints, psychologic distress, and treatment outcome in two groups of TMD patients, one previously subjected to whiplash injury. *J. Orofac. Pain* **1998**, *12*, 136–144. [PubMed]

29. Manfredini, D.; Favero, L.; Del Giudice, A.; Masiero, S.; Stellini, E.; Guarda-Nardini, L. Axis II psychosocial findings predict effectiveness of TMJ hyaluronic acid injections. *Int. J. Oral Maxillofac. Surg.* **2013**, *42*, 364–368. [CrossRef] [PubMed]

30. Nilsson, H.; Ekberg, E. Do psychological factors and general health influence the short-term efficacy of resilient appliance therapy in patients with temporomandibular disorder pain? *Acta Odontol. Scand.* **2010**, *68*, 141–147. [CrossRef] [PubMed]

31. Plaza-Manzano, G.; Delgado-de-la-Serna, P.; Díaz-Arribas, M.J.; Rodrigues-de-Souza, D.P.; Fernández-de-Las-Peñas, C.; Alburquerque-Sendín, F. Influence of Clinical, Physical, Psychological, and Psychophysical Variables on Treatment Outcomes in Somatic Tinnitus Associated with Temporomandibular Pain: Evidence From a Randomized Clinical Trial. *Pain Pract.* **2021**, *21*, 8–17. [CrossRef]

32. Abrahamsen, R.; Zachariae, R.; Svensson, P. Effect of hypnosis on oral function and psychological factors in temporomandibular disorders patients. *J. Oral Rehabil.* **2009**, *36*, 556–570. [CrossRef]

33. Fernandes, M.P.; Vidal, C.L.; de Oliveira, T.T.; Botelho, A.L.; Valente, M.; Dos Reis, A.C. Effects of photobiomodulation on auriculotherapy points for sleep disorders, anxiety, and temporomandibular dysfunctions. *Cranio* **2020**, 1–6. [CrossRef]

34. Iunes, D.H.; Chaves Éde, C.; Moura Cde, C.; Côrrea, B.; Carvalho, L.C.; Silva, A.M.; de Carvalho, E.C. Role of Auriculotherapy in the Treatment of Temporomandibular Disorders with Anxiety in University Students. *Evid. Based Complement. Alternat. Med.* **2015**, *2015*, 430143. [CrossRef]

35. Ivkovic, N.; Mladenovic, I.; Petkoci, S.; Stojic, D. TMD chronic pain and masseter silent period in psychiatric patients on antidepressive therapy. *J. Oral Rehabil.* **2008**, *35*, 424–432. [CrossRef]

36. Rodrigues, J.H.; Marques, M.M.; Biasotto-Gonzalez, D.A.; Moreira, M.S.; Bussadori, S.K.; Mesquita-Ferrari, R.A.; Martins, M.D. Evaluation of pain, jaw movements, and psychosocial factors in elderly individuals with temporomandibular disorder under laser phototherapy. *Lasers Med. Sci.* **2015**, *30*, 953–959. [CrossRef]

37. Brandão, R.; Mendes, C.M.C.; Brandão Filho, R.A.; De Sena, E.P. Isotonic exercises and relaxing techniques in individuals with temporomandibular dysfunction. *Cranio* **2022**, *40*, 199–206. [CrossRef]

38. Turner, J.A.; Mancl, L.; Aaron, L.A. Short- and long-term efficacy of brief cognitive-behavioral therapy for patients with chronic temporomandibular disorder pain: A randomized, controlled trial. *Pain* **2006**, *121*, 181–194. [CrossRef] [PubMed]

39. Rodrigues, M.D.F.; Rodrigues, M.L.; Bueno, K.S.; Aroca, J.P.; Camilotti, V.; Busato, M.C.A.; Mendonça, M.J. Effects of low-power laser auriculotherapy on the physical and emotional aspects in patients with temporomandibular disorders: A blind, randomized, controlled clinical trial. *Complement. Ther. Med.* **2019**, *42*, 340–346. [CrossRef] [PubMed]

40. De la Torre Canales, G.; Poluha, R.L.; Alvarez Pinzón, Y.N.; Rodrigues Conti, P.C.; Manfredini, D.; Sánchez-Ayala, A.; Rizzatti-Barbosa, C.M. Effects of Botulinum Toxin Type A on the Psychosocial Features of Myofascial Pain TMD Subjects: A Randomized Controlled Trial. *J. Oral Facial Pain Headache* **2021**, *35*, 288–296. [CrossRef] [PubMed]

41. Checherita, L.E.; Burlea, L.S.; Stamatin, O.; Manuc, D. Impact of Ibruprofen Medication and Prosthetic Esthetical Treatment in Elderly Patients with Temporal Mandibular Joint Disorders and Deppresion. *Rev. De Chim.* **2018**, *69*, 831–836. [CrossRef]

42. Moleirinho-Alves, P.M.M.; Almeida, A.; Exposto, F.G.; Oliveira, R.; Pezarat-Correia, P.L.C. Effects of therapeutic exercise and aerobic exercise programmes on pain, anxiety and oral health-related quality of life in patients with temporomandibular disorders. *J. Oral Rehabil.* **2021**, *48*, 1201–1209. [CrossRef] [PubMed]

43. Rudy, T.E.; Turk, D.C.; Kubinski, J.A.; Zaki, H.S. Differential treatment responses of TMD patients as a function of psychological characteristics. *Pain* **1995**, *61*, 103–112. [CrossRef]

44. Arola, H.M.; Nicholls, E.; Mallen, C.; Thomas, E. Self-reported pain interference and symptoms of anxiety and depression in community-dwelling older adults: Can a temporal relationship be determined? *Eur. J. Pain* **2010**, *14*, 966–971. [CrossRef] [PubMed]

45. Kindler, S.; Samietz, S.; Houshmand, M.; Grabe, H.J.; Bernhardt, O.; Biffar, R.; Kocher, T.; Meyer, G.; Völzke, H.; Metelmann, H.R.; et al. Depressive and anxiety symptoms as risk factors for temporomandibular joint pain: A prospective cohort study in the general population. *J. Pain* **2012**, *13*, 1188–1197. [CrossRef] [PubMed]

46. Bradburn, M.J.; Deeks, J.J.; Berlin, J.A.; Russell Localio, A. Much ado about nothing: A comparison of the performance of meta-analytical methods with rare events. *Stat. Med.* **2007**, *26*, 53–77. [CrossRef]

47. Themessl-Huber, M. Weak evidence supports the use of psychosocial interventions for chronic orofacial pain. *Evid. Based Dent.* **2012**, *13*, 58. [CrossRef]

Systematic Review

Lavage Volume of Arthrocentesis in the Management of Temporomandibular Disorders: A Systematic Review and Meta-Analysis

Hei Christopher Tsui [1], Chun Mo Lam [1], Yiu Yan Leung [2], Kar Yan Li [3], Natalie Sui Miu Wong [2] and Dion Tik Shun Li [2,*]

1 Faculty of Dentistry, The University of Hong Kong, Hong Kong, China
2 Oral and Maxillofacial Surgery, Faculty of Dentistry, The University of Hong Kong, Hong Kong, China
3 Clinical Research Centre, Faculty of Dentistry, The University of Hong Kong, Hong Kong, China
* Correspondence: diontsli@hku.hk

Abstract: The aim of this study was to investigate the most effective lavage volume of arthrocentesis in the management of temporomandibular disorders. A comprehensive electronic search, based on the PRISMA guidelines, was performed, which included a computer search with specific keywords, a reference list search and a manual search. The inclusion criteria were the following: a randomized controlled trial, at least 20 subjects who underwent arthrocentesis, mention of the irrigation materials used for the arthrocentesis, mention of the irrigation volumes used for the arthrocentesis, MMO and pain measured as VAS or NRS, were reported as outcome figures, mention of a specific diagnosis or signs and symptoms, and inclusion of the data on the MMO or VAS/NRS at 6-month follow-up. Sixteen publications were enrolled in the meta-analysis, comparing arthrocentesis with a lavage volume <150 mL and arthrocentesis with a lavage volume $\geq$150 mL, in the efficacy of the improvement in the mouth opening and pain reduction. The results revealed the group with a lavage volume <150 mL had a greater improvement in the mouth opening and pain reduction. However, results are to be interpreted with caution, due to the paucity of the randomized controlled literature and other confounding factors. Further high-quality studies are required to provide a better conclusion to the treatment outcomes of the different lavage volumes.

Keywords: arthrocentesis; temporomandibular joint disorders; temporomandibular joint; therapeutic irrigation

Citation: Tsui, H.C.; Lam, C.M.; Leung, Y.Y.; Li, K.Y.; Wong, N.S.M.; Li, D.T.S. Lavage Volume of Arthrocentesis in the Management of Temporomandibular Disorders: A Systematic Review and Meta-Analysis. *Diagnostics* **2022**, *12*, 2622. https://doi.org/10.3390/diagnostics12112622

Academic Editor: Luis Eduardo Almeida

Received: 30 August 2022
Accepted: 19 October 2022
Published: 28 October 2022

Publisher's Note: MDPI stays neutral with regard to jurisdictional claims in published maps and institutional affiliations.

1. Introduction

Temporomandibular disorders (TMDs) are a series of clinical problems which affect the temporomandibular joint (TMJ), masticatory muscles and associated structures [1]. Based on the diagnostic criteria for a TMD (DC/TMD), established in 2014, a TMD was diagnosed, based on the physical examination (Axis I) and assessment of the psychosocial status and pain-related disability (Axis II) [1]. In more detail, the DC/TMD Axis I includes: (a) muscle disorders, including myalgia, myofascial pain and myofascial pain with referral, etc., (b) intra-articular joint disorders, including disc displacement with or without the reduction or mouth opening limitation; (c) other articular conditions, including arthralgia.

TMDs are the second most common musculoskeletal problem following chronic lower back pain. Around 6–12% of the general population is thought to be affected by TMDs [2,3]. TMDs usually affect people between 20 to 40 years of age, and are more prevalent in females [4,5]. Some of the common signs and symptoms are facial pain, limited mouth opening and joint sounds. In the United States, the estimated management cost was about USD 4 billion per year [6]. Moreover, patients with TMDs used a broader range of services and hence, consumed more resources [7].

The management of TMDs focuses on alleviating pain or joint noises, restoring normal joint function and improving the overall quality of life. The first line approach involves the non-surgical treatment that includes a soft diet, pharmacotherapy, such as non-steroidal anti-inflammatory drugs (NSAIDs), occlusal splint therapy and physiotherapy [8]. In particular to the muscle-related TMD patients, the conservative approaches, such as physical therapy, laser therapy, occlusal splints and acupuncture were effective in pain reduction [9]. Other than the conservative modalities, pharmacologic agents are widely used for the treatment of mild and moderate TMD. Common drugs that improved TMD pain include NSAIDs, opioids, corticosteroids, antidepressants, anticonvulsants, antiepileptics, muscle relaxants, sedatives and hypnotics [10]. The intramuscular injections of botulinum toxin (BTX) had shown to relieve the muscle pain from TMDs. Intra-articular joint injections with corticosteroids, hyaluronic acid (HA) and platelet-rich plasma, also showed improvement in pain and functions [10]. Novel agents, such as ozone, were used as a topical gaseous therapy to the muscles and injective agents at the temporomandibular joint and demonstrated promising results in the muscle and the articular TMD [11].

Although the success rate of non-surgical treatments is approximately 70%, some patients do not respond well to these treatments. Patients who are refractory to non-surgical therapies and have high levels of pain and dysfunction, are suitable for surgical interventions, such as meniscectomy, disc repositioning and condylotomy [12]. These invasive procedures are often associated with surgical risks. Minimal invasive procedures, such as arthrocentesis, serve as an appropriate alternative to surgical intervention [13]. Indications of arthrocentesis, included patients with internal TMDs not responding to conservative treatment, patients with anterior disc displacement with or without reduction, disc adhesions, synovitis/capsulitis and degenerative osteoarthritis. A review article concluded that arthrocentesis is a highly effective approach when taking into account the notable clinical benefits and the small number of complications [13,14].

Arthrocentesis of the TMJ was first described by Nitzan and applied on patients with severe, limited mouth opening [15]. Arthrocentesis of the TMJ refers to the lavage of the upper joint space with saline, without visualizing the joint. Studies showed that arthrocentesis decreased pain, increased the maximal incisal opening and the follow-up showed the prolonged relief of symptoms [16]. This treatment utilized the pumping actions and hydraulic pressure to remove adhesions and inflammatory mediators, and widened the joint space [17,18]. Studies suggested that arthrocentesis reduced pain and the functional impairment rapidly, as an initial therapy, when compared to conventional treatment [19]. With a success rate of over 80% [20], and being less invasive than surgical interventions, arthrocentesis has become a common therapeutic intervention for patients with TMDs.

Recent studies have investigated the factors that determine the effectiveness of arthrocentesis, such as the needle technique, adjunctive treatment and lavage volume [21–23]. In particular, there is no consensus on the lavage volume in arthrocentesis, and the commonly adopted volume ranges from 50 mL to 300 mL. When arthrocentesis was introduced in the TMJ, approximately 200 mL of lactated Ringer's solution was used [15]. Studies have suggested that a smaller lavage volume was equally effective in washing the upper joint space of the TMJ [22,24]. One study suggested that a change in lavage volumes did not provide a statistical significance on the reduction of pain and the maximum mouth opening [22]. To the best of the authors' knowledge, no meta-analysis has compared the different lavage volumes on the effectiveness of arthrocentesis, in terms of the pain level and the maximum mouth opening. The aim of this study is to determine whether different lavage volumes will affect the treatment outcome of arthrocentesis, in relation to pain and mouth opening.

2. Materials and Methods

2.1. Protocol and Registration

The PRISMA 2020 statement [25] was taken as the reference in reporting this systematic review and meta-analysis, while this review was not registered in PROSPERO.

2.2. Eligibility Criteria

Criteria for selection was based on the PICOTS framework, as follows: 'P' (population): adult humans with a definitive clinical diagnosis or specific signs and symptoms of TMDs. 'I' (intervention): arthrocentesis, or lysis and lavage, with a clear indication of irrigation volumes and materials used. 'C' (comparison): control or adjunctive treatment including an occlusal splint or physiotherapy, or compared with arthrocentesis with intra-articular injections, or compared with arthrocentesis with different intra-articular injections, or compared with arthrocentesis on patients with different diagnoses of TMDs. 'O' (outcomes): the primary outcome is un-assisted/undefined, painless maximum mouth opening (MMO) in millimeters. The secondary outcome is pain intensity at rest, measured by the visual analogue scale (VAS) or the numerical rating scale (NRS). 'T' (time): all studies should have their follow-up period of at least 6 months. 'S' (study design): a randomized, controlled clinical trial.

The exclusion criteria were as follows: studies not in English, full text not available, non-adult human studies, studies not related to TMDs or where there was no mention of specific diagnoses or signs and symptoms, studies not using arthrocentesis as intervention, studies that included less than 20 patients undergoing arthrocentesis, studies not using the MMO and VAS as measuring outcomes, non-clinical studies, technical notes, case reports and case series.

2.3. Data Collection and Processing Strategy

The search was conducted in a total of three rounds. In the first round, an electronic search in PubMed, Cochrane Library, EMBASE (OVID), Scopus and Web of Science was performed, updated to December 5, 2020. The following search terms were used: ("Temporomandibular Joint" OR "Temporomandibular Joint Disorders" OR "Temporomandibular Joint Dysfunction Syndrome" OR "Craniomandibular Disorders" OR TMJ OR TMD or CMD) AND (Arthrocentesis OR "Temporomandibular Joint Arthrocentesis" OR lysis OR lavage). The publication date, language or publication status were not restricted. The articles obtained from the search term were imported into EndNote 20 and duplicates were removed with the software. The abstracts of the articles were then reviewed for eligibility. The full texts of the eligible studies after the first-round screening were obtained and imported into EndNote 20 and were included in the second-round screening.

In the second round, a manual search of the oral and maxillofacial surgery-related journals was performed in three relevant international journals: International Journal of Oral and Maxillofacial Surgery, the Journal of Oral and Maxillofacial Surgery, and the Journal of Cranio-Maxillofacial Surgery. Moreover, the reference lists of all identified studies from the first round and from the manual search were also scanned for relevant articles relating to the management of TMDs with arthrocentesis. The relevant articles after the first and second rounds of screening were included for the third-round evaluation.

In the third-round screening, the full texts of the included studies were evaluated, based on the following inclusion criteria: (1) randomized controlled trial; (2) at least 20 subjects who underwent arthrocentesis; (3) mention of the irrigation materials used for arthrocentesis; (4) mention of the irrigation volumes used for arthrocentesis; (5) the MMO and pain measured as VAS or NRS were reported as outcome figures; (6) mention of the specific diagnosis or signs and symptoms; (7) inclusion of the data on the MMO or VAS/NRS at 6-month follow-up. Articles after the third round of screening were considered eligible for inclusion for critical appraisal. A standard form was devised for the evaluation of inclusion/exclusion criteria of the studies screened.

2.4. Data Extraction Strategy

Data from the eligible studies were extracted using a standard data extraction sheet, specifically designed for this review. The data items that were extracted and analyzed were as follows: sample size, age at treatment, diagnosis of a TMD, change in the MMO in millimeters, change in pain measurement, time of follow-up, adjunctive procedure with arthrocentesis, such as an occlusal splint, intra-articular injections, irrigation materials and irrigation volumes.

2.5. Risk of Bias Analysis

With the revised Cochrane risk-of-bias tool (RoB 2) [26], two authors (H.C.T. and N.S.N.W.) critically appraised the eligible studies from the third round, for the risk of bias analysis, for verifying their strength in scientific evidence. Five domains were set for the appraisal, as follows: (1) randomization process; (2) deviations from the intended interventions; (3) missing outcome data; (4) measurement of the outcome; (5) selection of the reported result. When there was any discrepancy during the appraisal process between the two reviewers, the modulation was performed by the third reviewer (D.T.S.L.).

2.6. Summary Measures

For the continuous data, the weighted mean difference (WMD) was used to calculate the MMO (in millimeters) and pain (VAS from 0–10). Due to the various different follow-up time points in the different studies, the values at the 6-month follow-up or equivalent (i.e., 6 months, 24 or 26 weeks, or 180 days) were used for the meta-analysis. The VAS scales were standardized to a scale of 0–10. To obtain the mean and standard deviation (SD) in the studies where the range and median were given, the statistical formulas 5 and 16, in the article by Hozo et al. were used [27]. The postoperative SD values were assumed to be the same as the preoperative values when only the preoperative MMO and VAS were available. The values were excluded from the meta-analysis when the SDs of both the preoperative and postoperative values were not available.

2.7. Data Synthesis and Analysis

The extracted WMDs of the continuous data (MMO and VAS) were used in the meta-analysis. Forest plots were constructed using the random effects model with a 95% confidence interval (CI). Heterogeneity between the studies was evaluated using the Chi^2 with $p < 0.10$ or I^2 statistic of >50% [28]. In order to access whether the lavage volume had any effect on the clinical outcomes, we divided the included studies into two groups: (1) arthrocentesis with a lavage volume <150 mL; and (2) arthrocentesis with a lavage volume $\geq$150 mL. The MMO and VAS of the two groups were compared. The STATA (StataCorp. 2019. Stata Statistical Software: Release 16.0. College Station, TX, USA) was used for all statistical analyses.

2.8. Risk of Bias across Studies

Funnel plots were used to measure the publication bias, defined as the tendency to publish the results that are statistically or clinically significant. This method is deemed suitable, with more than 10 studies in the meta-analysis [28].

3. Results

3.1. Study Selection

PRISMA 2020 flow diagram of the study is presented in Figure 1. An electronic database search resulted in a total of 2648 articles. Then, 645 articles remained after the removal of the duplicates. Following an initial screening of the titles and abstracts, 420 articles were excluded, due to the irrelevant topics. A total of 225 articles were included in the second round search. In the second round search, a manual search from 2010 to 2020 and a reference list search from the included studies resulted in five additional articles. Two hundred and thirty articles were included in the third-round evaluation. Of those, 214 articles were excluded due to the failure of one or more of the inclusion criteria, mentioned above. Thus, a total of 16 studies were included in the meta-analysis.

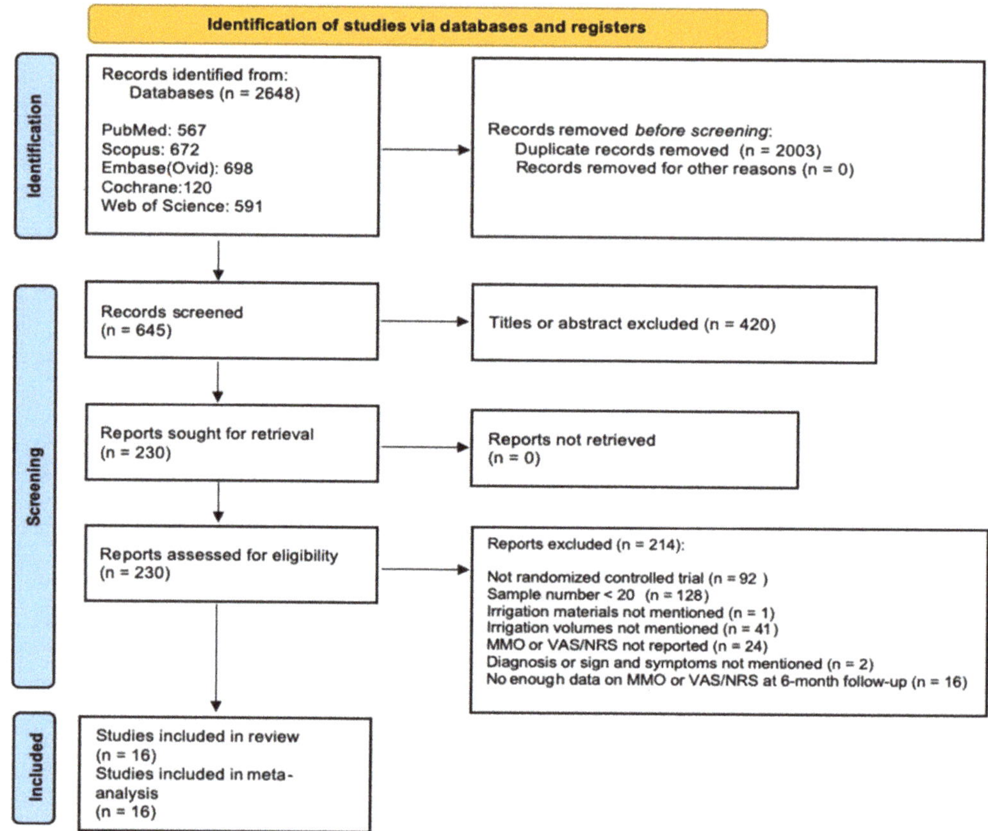

Figure 1. PRISMA 2020 flow diagram.

3.2. Study Characteristics

Details of the included studies are shown in Table 1. All studies are randomized control trials in the critical appraisal and meta-analysis. A total of 677 patients received arthrocentesis treatment for TMDs in the 16 included studies. In the included studies, the diagnoses included disc displacement with or without reduction (DDWR/DDWOR), Wilkes stages 3 and 4, internal derangement, osteoarthritis and arthralgia. In five of the included studies, less than 150 mL of lavage volume was used while eleven studies used more than or equal to 150 mL. Regarding the operative technique, 12 studies performed the traditional single-needle puncture while four other studies employed the double-needle technique. The intra-articular irrigants were saline and lactated Ringer's solution. Some studies adopted the adjunctive injection of hyaluronic acid (HA), dexamethasone, platelet-rich plasma (PRP) and bone marrow nucleated cells. The follow-up period in the included studies, ranged from immediately after the procedure to 24 months. For this review, the outcome data for 6 months, 24 or 26 weeks, or 180 days of follow-up were extracted.

Table 1. Details of the studies included.

Author	Year	Study Design	Subgroup	No. of Patients Who Underwent Arthrocentesis	Age at Treatment	Adjunctive Treatment	Irrigation Volume	(A: <150 mL, B: >150 mL)	Irrigation Material
Vos et al., (2014) [19]	2014	RCT	Arthrocentesis only	40	38.3		300 mL	B	NaCl
Guarda-Nardini et al., (2012) [23]	2012	RCT	SN grp	38	54.2		At least 300 + 1	B	Saline + HA
			TN grp (Control)	40	56.9			B	
Alpaslan et al., (2001) [29]	2001	RCT	Grp A (Control)	8	27		200–300	B	Saline
			Grp B	23			200–300 + 1	B	Saline + SH
Bayramoğlu et al., (2019) [30]	2019	RCT	SPA grp	16	25.9		100	A	Ringer's lactate
			DPA grp (Control)	16	25.75			A	
De Riu et al., (2019) [31]	2019	RCT	HA grp (Control)	15	44.5		200–250 + 2	B	Ringer's solution + SH
			BMNc grp	15	48.2		200–250 + 2	B	Ringer's solution + BMNc (Bone marrow nucleated cell)
Folle et al., (2018) [32]	2018	RCT	SPA grp	13	37.38		300 + 1	B	Saline + SH
			DPA grp (Control)	13	30.77			B	
Gorrela et al., (2016) [33]	2016	RCT	A grp (Control)	31	Not provided	Post op PT	100	A	Saline
			A + SH grp	31			100 + 1	A	Saline + SH
Huddleston Slater et al., (2012) [34]	2012	RCT	Group 1 (Control)	14	33.9	1 cc saline (placebo)	300 mL	B	Saline
			Group 2 (+dexamethasone)	14	32.6	1 cc Dexamethasone		B	
Murakami et al., (1995) [35]	1995	RCT	Group II: arthrocentesis	20	31.2		Nitzan's (2–3 mL Ringer's + 200 mL lactated Ringer's + 1 mL Celestone Soluspan)	B	Nitzan's (2–3 mL Ringer's + 200 mL lactated Ringer's + 1 mL Celestone Soluspan)
Patel et al., (2016) [36]	2016	RCT	Grp 1: Arthrocentesis only (Control)	15	Mean age not reported: 21–30 (43.33%)			B	Ringer's lactate
			Grp 2: Arthrocentesis + HA	15		Hyaluronic acid	2 mL (distend) + 200–300 mL	B	
Sipahi et al., (2015) [37]	2015	RCT	1 mL 5% Ringer's lactate (Control)	10	Mean age not reported: (16–50)		60–100 mL	A	Ringer's lactate
			Morphine 0.01 g made up to 10 mL Ringer's lactate	10				A	
			Tramadol 50 mg mixed with 5% Ringer's lactate 1 mL	10				A	
Tabrizi et al., (2014) [38]	2014	RCT	With Ringer only (Control)	30	28		2 mL saline (distend) + 200 mL	B	Ringer's lactate
			With Ringer + dexamethasone	30	27.07	With 8 mg dexamethasone		B	
Talaat et al., (2016) [39]	2016	RCT	Single needle	28	26.025	1 mL HA	300 mL	B	Saline
			Double needle (Control)	28				B	
Tatli et al., (2017) [40]	2017	RCT	Arthrocentesis only (Control)	40	35.2	2 mL HA	120 mL	A	NaCl
			Arthrocentesis + Stabilization splint	40	38.9			A	

Table 1. *Cont.*

Author	Year	Study Design	Subgroup	No. of Patients Who Underwent Arthrocentesis	Age at Treatment	Adjunctive Treatment	Irrigation Volume	(A: <150 mL, B: >150 mL)	Irrigation Material
Toameh et al., (2019) [41]	2019	RCT	Arthrocentesis only (Control)	10	40.53		5 mL (distend) + 100 mL	A	Ringer's lactate
			Arthrocentesis + HA	10	38.26	HA		A	
			Arthrocentesis + PRP	10	37.82	PRP		A	
Yapıcı-Yavuz et al., (2018) [42]	2018	RCT	Arthrocentesis + SH (diff in abstract and methods)	44	Not reported		Nitzan's (2–3 mL Ringer's + 200 mL lactated Ringer's + 1 mL Celestone Soluspan)	B	Nitzan's (2–3 mL Ringer's + 200 mL lactated Ringer's + 1 mL Celestone Soluspan)
			Arthrocentesis + methylpred-nisolone acetate					B	
			Arthrocentesis + tenoxicam					B	

3.3. Risk of Bias within the Studies

The assessment of the quality of the studies was carried out. Twelve studies showed a low risk of bias, three studies showed some concerns and one study showed a high risk of bias (Table 2).

Table 2. Risk of bias assessment.

Study ID	D1	D2	D3	D4	D5	Overall
Vos et al., (2014) [19]	+	+	+	+	+	+
Guarda-Nardini et al., (2012) [23]	+	+	+	+	+	+
Alpaslan et al., (2001) [29]	+	+	+	+	+	+
Bayramoğlu et al., (2019) [30]	+	+	+	+	+	+
De Riu et al., (2019) [31]	+	+	+	+	+	+
Folle et al., (2018) [32]	+	+	+	+	+	+
Gorrela et al., (2016) [33]	+	+	+	+	+	+
Huddleston Slater et al., (2012) [34]	+	+	+	+	+	+
Murakami et al., (1995) [35]	−	−	−	+	+	−
Patel et al., (2016) [36]	+	+	+	+	+	+
Sipahi et al., (2015) [37]	+	+	+	+	+	+
Tabrizi et al., (2014) [38]	+	+	+	+	+	+
Talaat et al., (2016) [39]	+	+	!	+	+	!

Table 2. *Cont.*

Study ID	D1	D2	D3	D4	D5	Overall
Tatli et al., (2017) [40]	+	+	+	+	+	+
Toameh et al., (2019) [41]	!	+	+	+	+	!
Yapıcı-Yavuz et al., (2018) [42]	!	+	+	+	+	!

+ Low risk; ! Some concerns; − High risk; D1: Randomisation process; D2: Deviations from the intended interventions; D3: Missing outcome data; D4: Measurement of the outcome; D5: Selection of the reported result.

3.4. Synthesis of the Results

The WMD and SD of the continuous variables were used for the meta-analyses. In one of the studies where the range and median were given [32], the mean and SD were estimated with statistical formulae [27]. In one study [34], the SD was calculated from the confidence intervals [28]. In two studies, the SD for the preoperative and postoperative pain measurements (VAS) were not available [29,30], therefore, the pain measurements from those studies were excluded from the meta-analysis. In three studies, the SD values of the postoperative MMO and VAS were not available. The SD of the postoperative MMO and VAS were assumed to be the same as the pre-operative values.

Random effects model was used for the construction of the forest plots, due to the intention of the generalization inference and the substantial heterogeneity found in the included studies (I^2 ranged from 70.02% to 98.5%; all $p < 0.1$).

3.5. MMO

The forest plots of the pooled WMD, between the <150 mL and ≥150 mL groups, in the improvement of the MMO, are summarized in Figure 2. All included studies demonstrated an improvement in the MMO after arthrocentesis (range 1.85–16.14 mm), with a greater improvement in the MMO in the group with a lavage volume <150 mL. There is no statistically significant difference in the MMO after the 6 months or equivalent (WMD: 9.62, 95% CI: 6.17 to 13.07, I^2 = 99.1%, p = 0.392) follow-up between the two groups.

A sensitivity analysis is not performed as there is only one study with high risk of bias.

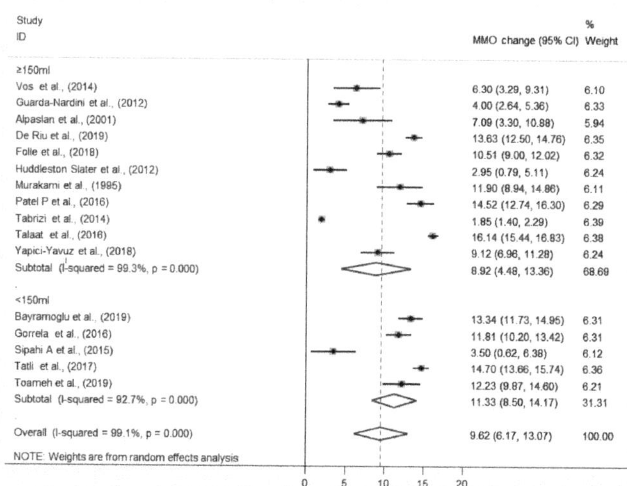

Figure 2. Forest plot using the random effects models showing the weighted mean difference (WMD) in the maximal mouth opening (MMO) between the two groups: all studies included. (Heterogeneity between the studies quantified using the I^2 and *p*-value of the Chi^2 test were listed) [19,23,29–42].

3.6. Pain (VAS)

The forest plots of the pooled WMD, between the <150 mL and ≥150 mL groups, in the reduction in VAS, are summarized in Figure 3. All included studies demonstrated a reduction in the VAS after arthrocentesis (range −1.23––8.20), with a greater improvement in the VAS in the group with a lavage volume <150 mL. There is no statistically significant difference in the VAS after the 6 months or equivalent (WMD: −4.91, 95% CI: −3.89 to −5.93, I^2 = 97.9%, *p* = 0.696) follow-up between the two groups.

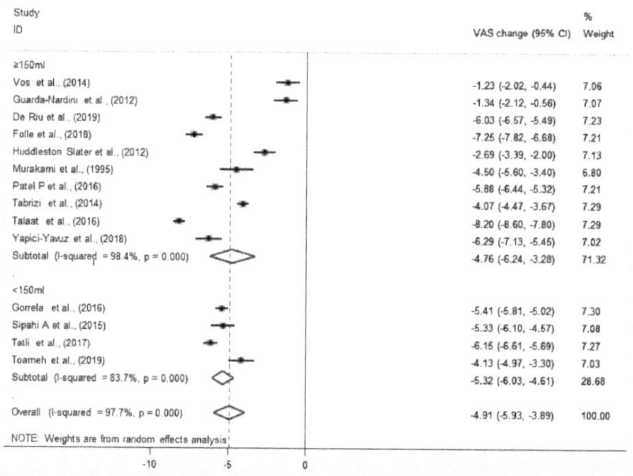

Figure 3. Forest plot using the random effects models showing the weighted mean difference (WMD) in the VAS between the two groups: all studies included. (Heterogeneity between the studies quantified using I^2 and the *p*-value of the Chi^2 test were listed) [19,23,31–42].

A sensitivity analysis is not performed as there is only one study with a high risk of bias.

3.7. Assessment of the Publication Bias

The publication bias was assessed using the funnel plot techniques and Begg's rank test. The funnel plots of the MMO and VAS mean differences were both not in a severe asymmetry, which are shown in Figures 4 and 5. These implied the small study effect and thus a publication bias was not significant. Begg's rank test also suggests no significant publication bias in the MMO ($p = 0.163$ in overall and 0.640 and 0.086 in the subgroups) and the VAS differences ($p = 0.155$ in overall and 0.210 and 0.734 in the subgroups).

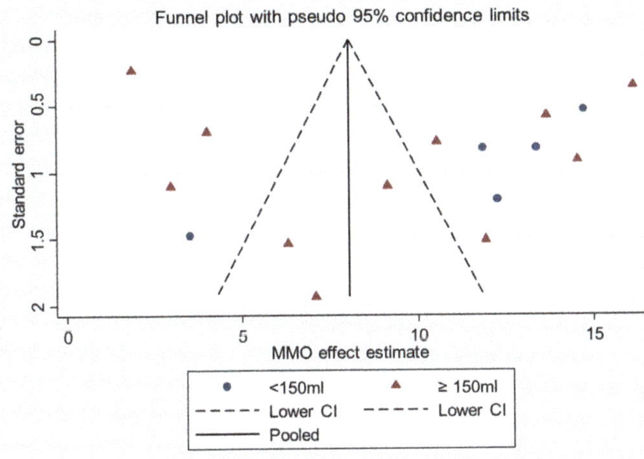

Figure 4. Funnel plot of the MMO mean differences.

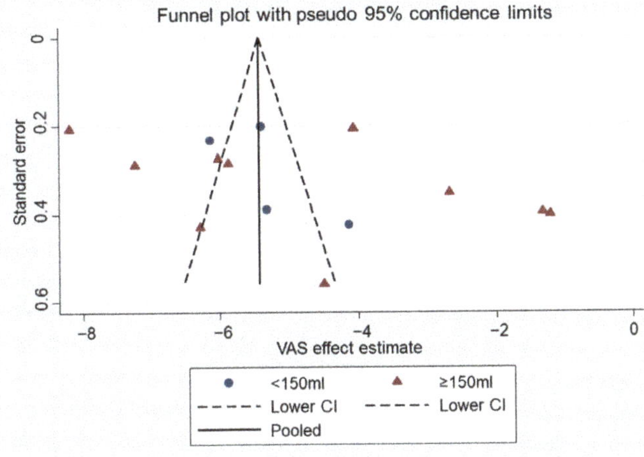

Figure 5. Funnel plot of the VAS mean differences.

4. Discussion

In the management of arthrogenous TMDs, arthrocentesis has become a standard treatment option, due to its high efficacy and safety [14,43]. However, different techniques of arthrocentesis exist in the literature and in clinical practice, such as additional injection materials into the superior joint space [41,42,44–48], the use of ultra-sound guidance [49–52] single versus double puncture techniques [30,32,53,54] and the timing of the procedure [19,55]. Specifically, the ideal irrigation volume for arthrocentesis of the TMJ remains a controversy. For example, in the studies included in this review, the irrigation volume ranged from 60 mL to 301 mL. Studies regarding the ideal irrigation volume for the TMJ arthrocentesis are few and far between, let alone the prospective clinical trials. If arthrocentesis performed with a smaller irrigation volume results in similar clinical outcomes, then arthrocentesis with a larger irrigation volume would be unnecessary, and the procedure could be completed in a timely fashion with increased patient comfort. Unfortunately, such a recommendation is difficult to make, due to the obvious knowledge gap at present. To the authors' knowledge, the present study is the first systematic review and meta-analysis to investigate the ideal irrigation volume for the TMJ arthrocentesis.

The results of our study suggested that arthrocentesis with a smaller irrigation volume (<150 mL) may be superior to that performed with a larger irrigation volume ($\geq$150 mL), in terms of pain reduction and jaw function. Our results showed that there was a greater improvement in the MMO and VAS in the group with a small irrigation volume (<150 mL). Nevertheless, the evidence may be inconclusive as there was no statistically significant difference between the smaller (<150 mL) and larger ($\geq$150 mL) irrigation volumes in the MMO (WMD: 9.62, 95% CI: 6.17 to 218 13.07, I^2 = 99.1%, p = 0.392) and VAS (WMD: -4.91, 95% CI: -3.89 to -5.93, I^2 = 236 97.9%, p = 0.696). The pooled analyses represented by the forest plots are, however, from a limited number of studies. Although whether such difference seen in this study represents any actual clinical significance, is unknown, it may be safe to propose that arthrocentesis with a smaller volume (<150 mL), is at least as effective as that performed with a larger volume. However, since only the data from the 6-month follow-up was extracted from the included studies for the meta-analysis, it is not possible to say whether this is also true in the longer term.

The view that a smaller volume used for arthrocentesis of the TMJ may be just as effective as a larger volume from the current study, is shared with the few studies in the current literature on the topic. In a clinical study by Grossman et al., arthrocentesis of the TMJ was used to treat patients presented with disc displacement without reduction [22]. The patients were divided into two groups, with 50 mL or 200 mL of irrigation volume used in the procedure. While it was found that the favorable clinical outcomes were seen in both groups, in terms of pain reduction and improvement of the jaw function, no significant difference was found between the two groups. In another randomized controlled trial in 2017, no statistical significant difference was found between the groups with an irrigation volume of either 100 mL or 250 mL, while the clinical improvement of the TMD symptoms were seen in both groups [24]. Moreover, in a recent cadaveric study, it was found that 25 mL of the irrigation solution was sufficient to remove methylene blue from the TMJ space of fresh human cadavers [56]. In contrast, in a clinical trial by Kaneyama et al., it was suggested that the ideal lavage volume for the removal of inflammatory mediators from the TMJ joint space, was between 300–400 mL, although the statistical analysis of the clinical variables of the TMDs, such as pain reduction and mouth opening, were not reported in that study [57]. Therefore, more clinical studies are required in order to further understand the ideal irrigation volume for TMJ arthrocentesis.

There were a number of limitations to the current study. Due to the paucity of randomized controlled studies in the literature, it was not possible to compare the results of those studies performed with control groups that investigated the effect of different irrigation volumes. Rather, the results of the randomized controlled trials, which have reported the irrigation volume and outcomes of arthrocentesis and not focusing on the irrigation volumes, were pooled for our meta-analysis. This methodology is not ideal and therefore the results from this study cannot be interpreted with high certainty. Moreover, confounding factors were present across the studies, such as different diagnoses of TMDs, different materials used for the lavage and intra-articular injection, different techniques of arthrocentesis across centers and the difference in the mean age and sex distribution in the included studies.

The association of ethnic background and prevalence of TMD were discussed in a number of studies [58–60]. It has been a less discussed factor and deserved more discussion in the context of the arthrocentesis outcome. Nevertheless, due to a limited number of studies specifying the patients' ethnicity in the arthrocentesis treatment, a comprehensive systematic review becomes difficult. Other factors, such as age, duration of the symptoms and oral habits are believed to affect the prognosis of arthrocentesis [61]. Therefore, ethnicity can be investigated as an effect modifier in the arthrocentesis when a different lavage volume is applied. A more in depth understanding of the relationship between ethnicity, age, gender and different irrigation volumes of arthrocentesis is feasible when a diverse population is included in further studies.

Moreover, only 6-months of data were selected for the meta-analysis, the effects of the different irrigation volume at different time points are unknown. Further review on the impact of the irrigation volume can be analyzed according to the duration of the follow-up. In fact, one study had identified the impact of the follow-up time of the different treatment for TMDs, by performing sub-group analyses, according to the duration of follow-up [14]. Therefore, more randomized controlled trials of arthrocentesis involving different follow-up periods are required before a final conclusion can be drawn.

5. Conclusions

In conclusion, the current systematic review and meta-analysis suggests that arthrocentesis of the TMJ is at least as effective, if not more, when a smaller lavage volume is used (<150 mL). However, due to the limitations with the methodology and confounding factors, the evidence is weak at this time. Future randomized clinical trials are needed to better understand the clinical outcomes related to the different irrigation volumes used for the arthrocentesis of the TMJ.

Author Contributions: Conceptualization, D.T.S.L., H.C.T. and C.M.L.; methodology, D.T.S.L., H.C.T. and C.M.L.; formal analysis, H.C.T., C.M.L. and K.Y.L.; writing—original draft preparation, D.T.S.L., H.C.T. and C.M.L.; writing—review and editing, Y.Y.L., D.T.S.L. and N.S.M.W.; supervision, D.T.S.L. All authors have read and agreed to the published version of the manuscript.

Funding: This research received no external funding.

Institutional Review Board Statement: Not applicable.

Informed Consent Statement: Not applicable.

Data Availability Statement: Data is available in the manuscript.

Conflicts of Interest: The authors declare no conflict of interest.

References

1. Schiffman, E.; Ohrbach, R.; Truelove, E.; Look, J.; Anderson, G.; Goulet, J.P.; List, T.; Svensson, P.; Gonzalez, Y.; Lobbezoo, F.; et al. Diagnostic Criteria for Temporomandibular Disorders (DC/TMD) for clinical and research applications: Recommendations of the international RDC/TMD consortium network and orofacial pain special interest group. *J. Oral Facial Pain Headache* **2014**, *28*, 6–27. [CrossRef]
2. Lipton, J.A.; Ship, J.A.; Larach-Robinson, D. Estimated prevalence and distribution of reported orofacial pain in the United States. *J. Am. Dent. Assoc.* **1993**, *124*, 115–121. [CrossRef] [PubMed]
3. Magnusson, T.; Egermark, I.; Carlsson, G.E. A Longitudinal Epidemiologic Study of Signs and Symptoms of Temporomandibular Disorders from 15 to 35 Years of Age. *J. Orofac. Pain* **2000**, *14*, 310–319. [PubMed]
4. Manfredini, D.D.D.S.; Guarda-Nardini, L.M.D.D.S.; Winocur, E.D.M.D.; Piccotti, F.D.D.S.; Ahlberg, J.D.D.S.P.; Lobbezoo, F.D.D.S.P. Research diagnostic criteria for temporomandibular disorders: A systematic review of axis I epidemiologic findings. *Oral Surg. Oral Med. Oral Pathol. Oral Radiol. Endodontology* **2011**, *112*, 453–462. [CrossRef]
5. Tore Gran, J. The epidemiology of chronic generalized musculoskeletal pain. *Best Pract. Res. Clin. Rheumatol.* **2003**, *17*, 547–561. [CrossRef]
6. Facial Pain. Available online: https://www.nidcr.nih.gov/research/data-statistics/facial-pain (accessed on 22 December 2021).

7. Alex White, B.; Williams, L.A.; Leben, J.R. Health Care Utilization and Cost among Health Maintenance Organization Members with Temporomandibular Disorders. *J. Orofac. Pain* **2001**, *15*, 158–169.

8. Dimitroulis, G. Management of temporomandibular joint disorders: A surgeon's perspective. *Aust. Dent. J.* **2018**, *63*, S79–S90. [CrossRef] [PubMed]

9. Ferrillo, M.; Ammendolia, A.; Paduano, S.; Calafiore, D.; Marotta, N.; Migliario, M.; Fortunato, L.; Giudice, A.; Michelotti, A.; de Sire, A. Efficacy of Rehabilitation on Reducing Pain in Muscle-Related Temporomandibular Disorders: A Systematic Review and Meta-Analysis of Randomized Controlled Trials. *J. Back Musculoskelet. Rehabil.* **2022**, *35*, 921–936. [CrossRef]

10. Andre, A.; Kang, J.; Dym, H. Pharmacologic Treatment for Temporomandibular and Temporomandibular Joint Disorders. *Oral Maxillofac. Surg. Clin. North Am.* **2022**, *34*, 49–59. [CrossRef]

11. De Sire, A.; Marotta, N.; Ferrillo, M.; Agostini, F.; Sconza, C.; Lippi, L.; Respizzi, S.; Giudice, A.; Invernizzi, M.; Ammendolia, A. Oxygen-Ozone Therapy for Reducing pro-Inflammatory Cytokines Serum Levels in Musculoskeletal and Temporomandibular Disorders: A Comprehensive Review. *Int. J. Mol. Sci.* **2022**, *23*, 2528. [CrossRef]

12. Barkin, S.; Weinberg, S. Internal derangements of the temporomandibular joint: The role of arthroscopic surgery and arthrocentesis. *J. Can. Dent. Assoc.* **2000**, *66*, 199–203. [PubMed]

13. Soni, A. Arthrocentesis of Temporomandibular Joint- Bridging the Gap between Non-Surgical and Surgical Treatment. *Ann. Maxillofac. Surg.* **2019**, *9*, 158. [CrossRef]

14. Al-Moraissi, E.A.; Wolford, L.M.; Ellis, E.; Neff, A. The Hierarchy of Different Treatments for Arthrogenous Temporomandibular Disorders: A Network Meta-Analysis of Randomized Clinical Trials. *J. Cranio-Maxillofac. Surg.* **2020**, *48*, 9–23. [CrossRef] [PubMed]

15. Nitzan, D.W.; Dolwick, M.F.; Martinez, G.A. Temporomandibular joint arthrocentesis: A simplified treatment for severe limited mouth opening. *Plast. Reconstr. Surg.* **1992**, *90*, 1132. [CrossRef]

16. Trieger, N.; Hoffman, C.H.; Rodriguez, E. The effect of arthrocentesis of the temporomandibular joint in patients with rheumatoid arthritis. *J. Oral Maxillofac. Surg.* **1999**, *57*, 537–540. [CrossRef]

17. Alpaslan, C.; Bilgihan, A.; Alpaslan, G.H.; Güner, B.; Yis, M.Ö.; Erbaş, D. Effect of arthrocentesis and sodium hyaluronate injection on nitrite, nitrate, and thiobarbituric acid-reactive substance levels in the synovial fluid. *Oral Surg. Oral Med. Oral Pathol. Oral Radiol. Endodontology* **2000**, *89*, 686–690. [CrossRef]

18. Yura, S.; Totsuka, Y. Relationship between effectiveness of arthrocentesis under sufficient pressure and conditions of the temporomandibular joint. *J. Oral Maxillofac. Surg.* **2005**, *63*, 225–228. [CrossRef]

19. Vos, L.M.; Huddleston Slater, J.J.R.; Stegenga, B. Arthrocentesis as initial treatment for temporomandibular joint arthropathy: A randomized controlled trial. *J. Cranio-Maxillofac. Surg.* **2013**, *42*, e134–e139. [CrossRef]

20. Monje-Gil, F.; Nitzan, D.; González-Garcia, R. Temporomandibular joint arthrocentesis. Review of the literature. *Med. Oral Patol. Oral Cir. Bucal.* **2012**, *17*, e575–e581. [CrossRef]

21. Bas, B.; Kazan, D.; Kutuk, N.; Gurbanov, V. The Effect of Exercise on Range of Movement and Pain after Temporomandibular Joint Arthrocentesis. *J. Oral Maxillofac. Surg.* **2018**, *76*, 1181–1186. [CrossRef]

22. Grossmann, E.; Poluha, R.L.; Iwaki, L.C.V.; Iwaki Filho, L. Arthrocentesis with different irrigation volumes in patients with disc displacement without reduction: One-year follow-up. *CRANIO®* **2020**, *38*, 122–127. [CrossRef] [PubMed]

23. Guarda-Nardini, L.; Ferronato, G.; Manfredini, D. Two-needle vs. single-needle technique for TMJ arthrocentesis plus hyaluronic acid injections: A comparative trial over a six-month follow up. *Int. J. Oral Maxillofac. Surg.* **2012**, *41*, 506–513. [CrossRef] [PubMed]

24. De Barros Melo, M.; Dos Santos Melo, J.; Sarmento, V.; De Azevedo, R.; Queiroz, C. Influence of arthrocentesis irrigation volume at temporomandibular disorder treatment. *Indian J. Dent. Res.* **2017**, *28*, 655–660.

25. Page, M.J.; McKenzie, J.; Bossuyt, P.; Boutron, I.; Hoffmann, T.; Mulrow, C.; Shamseer, L.; Tetzlaff, J.; Akl, E.; Brennan, S.E.; et al. The Prisma 2020 Statement: An Updated Guideline for Reporting Systematic Reviews. *Syst. Rev.* **2020**, *10*, 89. [CrossRef]

26. Sterne, J.A.; Savović, J.; Page, M.J.; Elbers, R.G.; Blencowe, N.S.; Boutron, I.; Cates, C.J.; Cheng, H.-Y.; Corbett, M.S.; Eldridge, S.M.; et al. Rob 2: A Revised Tool for Assessing Risk of Bias in Randomised Trials. *BMJ* **2019**, *366*, l4898. [CrossRef] [PubMed]

27. Hozo, S.P.; Djulbegovic, B.; Hozo, I. Estimating the mean and variance from the median, range, and the size of a sample. *BMC Med. Res. Methodol.* **2005**, *5*, 13. [CrossRef]

28. Higgins, J.P.T.; Green, S. (Eds.) *Cochrane Handbook for Systematic Reviews of Interventions*; Version 5.1.0.; John Wiley & Sons: Hoboken, NJ, USA, 2011.

29. Alpaslan, G.H.; Alpaslan, C. Efficacy of temporomandibular joint arthrocentesis with and without injection of sodium hyaluronate in treatment of internal derangements. *J. Oral Maxillofac. Surg.* **2001**, *59*, 613–618. [CrossRef]

30. Bayramoğlu, Z.; Tozoğlu, S. Comparison of single- and double-puncture arthrocentesis for the treatment of temporomandibular joint disorders: A six-month, prospective study. *CRANIO®* **2019**, *39*, 151–156. [CrossRef]

31. De Riu, G.; Vaira, L.A.; Carta, E.; Meloni, S.M.; Sembronio, S.; Robiony, M. Bone Marrow Nucleated Cell Concentrate Autograft in Temporomandibular Joint Degenerative Disorders: 1-Year Results of a Randomized Clinical Trial. *J. Cranio-Maxillofac. Surg.* **2019**, *47*, 1728–1738. [CrossRef]

32. Folle, F.S.; Poluha, R.L.; Setogutti, E.T.; Grossmann, E. Double puncture versus single puncture arthrocentesis for the management of unilateral temporomandibular joint disc displacement without reduction: A randomized controlled trial. *J. Cranio-Maxillofac. Surg.* **2018**, *46*, 2003–2007. [CrossRef]

33. Gorrela, H.; Prameela, J.; Srinivas, G.; Reddy, B.V.; Sudhir, M.V.S.; Arakeri, G. Efficacy of Temporomandibular Joint Arthrocentesis with Sodium Hyaluronate in the Management of Temporomandibular Joint Disorders: A Prospective Randomized Control Trial. *J. Maxillofac. Oral Surg.* **2016**, *16*, 479–484. [CrossRef] [PubMed]

34. Huddleston-Slater, J.J.R.; Vos, L.M.; Stroy, L.P.P.; Stegenga, B. Randomized Trial on the Effectiveness of Dexamethasone in TMJ Arthrocentesis. *J. Dent. Res.* **2012**, *91*, 173–178. [CrossRef] [PubMed]

35. Murakami, K.I.; Hosaka, H.; Moriya, Y.; Segami, N.; Iizuka, T. Short-Term Treatment Outcome Study for the Management of Temporomandibular Joint Closed Lock. *Oral Surg. Oral Med. Oral Pathol. Oral Radiol. Endodontology* **1995**, *80*, 253–257. [CrossRef]

36. Patel, P.; Idrees, F.; Newaskar, V.; Agrawal, D. Sodium Hyaluronate: An Effective Adjunct in Temporomandibular Joint Arthrocentesis. *Oral Maxillofac. Surg.* **2016**, *20*, 405–410. [CrossRef] [PubMed]

37. Sipahi, A.; Satilmis, T.; Basa, S. Comparative Study in Patients with Symptomatic Internal Derangements of the Temporomandibular Joint: Analgesic Outcomes of Arthrocentesis with or without Intra-Articular Morphine and Tramadol. *Br. J. Oral Maxillofac. Surg.* **2015**, *53*, 316–320. [CrossRef]

38. Tabrizi, R.; Karagah, T.; Arabion, H.; Soleimanpour, M.R.; Soleimanpour, M. Outcomes of Arthrocentesis for the Treatment of Internal Derangement Pain. *J. Craniofac. Surg.* **2014**, *25*, e571–e575. [CrossRef]

39. Talaat, W.; Ghoneim, M.M.; Elsholkamy, M. Single-Needle Arthrocentesis (Shepard Cannula) vs. Double-Needle Arthrocentesis for Treating Disc Displacement without Reduction. *CRANIO®* **2016**, *34*, 296–302. [CrossRef]

40. Tatli, U.; Benlidayi, M.E.; Ekren, O.; Salimov, F. Comparison of the Effectiveness of Three Different Treatment Methods for Temporomandibular Joint Disc Displacement without Reduction. *Int. J. Oral Maxillofac. Surg.* **2017**, *46*, 603–609. [CrossRef]

41. Toameh, M.H.; Alkhouri, I.; Karman, M.A. Management of patients with disk displacement without reduction of the temporomandibular joint by arthrocentesis alone, plus hyaluronic acid or plus platelet-rich plasma. *Dent. Med. Probl.* **2019**, *56*, 265–272. [CrossRef]

42. Yapıcı-Yavuz, G.; Şimşek-Kaya, G.; Oğul, H. A comparison of the effects of methylprednisolone acetate, sodium hyaluronate and tenoxicam in the treatment of non-reducing disc displacement of the temporomandibular joint. *Med. Oral Patol. Oral Cir. Bucal.* **2018**, *23*, e351–e358.

43. Polat, M.E.; Yanik, S. Efficiency of arthrocentesis treatment for different temporomandibular joint disorders. *Int. J. Oral Maxillofac. Surg.* **2020**, *49*, 621–627. [CrossRef] [PubMed]

44. Bouloux, G.F.; Chou, J.; Krishnan, D.; Aghaloo, T.; Kahenasa, N.; Smith, J.A.; Giannakopoulos, H. Is Hyaluronic Acid or Corticosteroid Superior to Lactated Ringer Solution in the Short Term for Improving Function and Quality of Life after Arthrocentesis? Part 2. *J. Oral Maxillofac. Surg.* **2017**, *75*, 63–72. [CrossRef] [PubMed]

45. Haigler, M.C.; Abdulrehman, E.; Siddappa, S.; Kishore, R.; Padilla, M.; Enciso, R. Use of platelet-rich plasma, platelet-rich growth factor with arthrocentesis or arthroscopy to treat temporomandibular joint osteoarthritis: Systematic review with meta-analyses. *J. Am. Dent. Assoc.* **2018**, *149*, 940–952.e942. [CrossRef] [PubMed]

46. Kiliç, S.C. Does Injection of Corticosteroid after Arthrocentesis Improve Outcomes of Temporomandibular Joint Osteoarthritis? A Randomized Clinical Trial. *J. Oral Maxillofac. Surg.* **2016**, *74*, 2151–2158. [CrossRef]

47. Su, N.; Yang, X.; Liu, Y.; Huang, Y.; Shi, Z. Evaluation of arthrocentesis with hyaluronic acid injection plus oral glucosamine hydrochloride for temporomandibular joint osteoarthritis in oral-health-related quality of life. *J. Cranio-Maxillofac. Surg.* **2013**, *42*, 846–851. [CrossRef] [PubMed]

48. Yilmaz, O.; Korkmaz, Y.T.; Tuzuner, T. Comparison of treatment efficacy between hyaluronic acid and arthrocentesis plus hyaluronic acid in internal derangements of temporomandibular joint. *J. Cranio-Maxillofac. Surg.* **2019**, *47*, 1720–1727. [CrossRef]

49. Bhargava, D.; Thomas, S.; Pawar, P.; Jain, M.; Pathak, P. Ultrasound-guided arthrocentesis using single-puncture, double-lumen, single-barrel needle for patients with temporomandibular joint acute closed lock internal derangement. *Oral Maxillofac. Surg.* **2019**, *23*, 159–165. [CrossRef] [PubMed]

50. Bilgir, E.; Yıldırım, D.; Şenturk, M.F.; Orhan, H. Clinical and ultrasonographic evaluation of ultrasound-guided single puncture temporomandibular joint arthrocentesis. *Oral Maxillofac. Surg.* **2020**, 1–10. [CrossRef]

51. Hu, Y.; Zhang, X.; Liu, S.; Xu, F. Ultrasound-guided vs. conventional arthrocentesis for management of temporomandibular joint disorders: A systematic review and meta-analysis. *CRANIO®* **2020**, 1–10. [CrossRef]

52. Leung, Y.Y.; Wu, F.H.W.; Chan, H.H. Ultrasonography-guided arthrocentesis versus conventional arthrocentesis in treating internal derangement of temporomandibular joint: A systematic review. *Clin. Oral Investig.* **2020**, *24*, 3771–3780. [CrossRef]

53. Monteiro, J.L.G.C.; de Arruda, J.A.A.; Silva, E.D.d.O.e.; Vasconcelos, B.C.d.E. Is Single-Puncture TMJ Arthrocentesis Superior to the Double-Puncture Technique for the Improvement of Outcomes in Patients With TMDs? *J. Oral Maxillofac. Surg.* **2020**, *78*, 1319-e1. [CrossRef] [PubMed]

54. Nagori, S.A.; Roy Chowdhury, S.K.; Thukral, H.; Jose, A.; Roychoudhury, A. Single puncture versus standard double needle arthrocentesis for the management of temporomandibular joint disorders: A systematic review. *J. Oral Rehabil.* **2018**, *45*, 810–818. [CrossRef] [PubMed]

55. Li, D.T.S.; Wong, N.S.M.; Li, S.K.Y.; McGrath, C.P.; Leung, Y.Y. Timing of arthrocentesis in the management of temporomandibular disorders: An integrative review and meta-analysis. *Int. J. Oral Maxillofac. Surg.* **2021**, *50*, 1078–1088. [CrossRef] [PubMed]

56. Palma, L.F.; Rocha, P.R.; Chaddad Neto, F.E.A.; Smith, R.L.; de Moraes, L.O.C. Irrigation fluid volume requirement for conventional arthrocentesis of the temporomandibular joint: A cadaver study. *Int. J. Oral Maxillofac. Surg.* **2020**, *49*, 1459–1463. [CrossRef] [PubMed]

57. Kaneyama, K.; Segami, N.; Nishimura, M.; Sato, J.; Fujimura, K.; Yoshimura, H. The ideal lavage volume for removing bradykinin, interleukin-6, and protein from the temporomandibular joint by arthrocentesis. *J. Oral Maxillofac. Surg.* **2004**, *62*, 657–661. [CrossRef] [PubMed]

58. Plesh, O.; Adams, S.H.; Gansky, S.A. Racial/Ethnic and Gender Prevalences in Reported Common Pains in a National Sample. *J. Orofac. Pain* **2011**, *25*, 25–31. [PubMed]

59. Lerman, S.F.; Campbell, C.M.; Buenaver, L.F.; Medak, M.; Phillips, J.; Polley, M.; Smith, M.T.; Haythornthwaite, J.A. Exploring the Role of Negative Cognitions in the Relationship between Ethnicity, Sleep, and Pain in Women with Temporomandibular Joint Disorder. *J. Pain* **2018**, *19*, 1342–1351. [CrossRef] [PubMed]

60. Sipilä, K.; Tolvanen, M.; Mitrirattanakul, S.; Sitthisomwong, P.; Järvelin, M.-R.; Taanila, A.; Anttonen, V.; Lahti, S. Orofacial Pain and Symptoms of Temporomandibular Disorders in Finnish and Thai Populations. *Acta Odontol. Scand.* **2015**, *73*, 330–335. [CrossRef]

61. Kim, Y.H.; Jeong, T.M.; Pang, K.M.; Song, S.I. Influencing factor on the prognosis of arthrocentesis. *J. Korean Assoc. Oral Maxillofac. Surg.* **2014**, *40*, 155–159. [CrossRef]

MDPI

St. Alban-Anlage 66

4052 Basel

Switzerland

www.mdpi.com

Diagnostics Editorial Office

E-mail: diagnostics@mdpi.com

www.mdpi.com/journal/diagnostics